DEDICATION

THE LAST MEETING of the Panel on Economic, Environmental, and Social Outcomes of Dam Removal was held in Washington, D.C., on September 11–12, 2001. The panel was in the midst of a discussion of the Manatawny Creek dam removal project when the first attack took place on the World Trade Center in New York City. As that morning unfolded, we learned of the second attack in New York, the attack on the Pentagon, and the plane crash in Pennsylvania. None of us will forget where we were on September 11, 2001, nor will we forget the thousands of lives lost as a result of such senseless and brutal acts. We dedicate this report to the victims and their families and to the courageous firefighters, police, and rescue teams in New York, Washington, D.C., and Pennsylvania.

THE H. JOHN HEINZ III CENTER FOR SCIENCE, ECONOMICS AND THE ENVIRONMENT

Board of Trustees

Sustainable Oceans, Coasts, and Waterways Steering Committee

DAM REMOVAL

Science and Decision Making

THE
HEINZ
CENTER

THE H. JOHN HEINZ III CENTER FOR
SCIENCE, ECONOMICS AND THE ENVIRONMENT

The H. John Heinz III Center for Science, Economics and the Environment

Established in December 1995 in honor of Senator John Heinz, The Heinz Center is a nonprofit institution dedicated to improving the scientific and economic foundation for environmental policy through multisectoral collaboration. Focusing on issues that are likely to confront policymakers within two to five years, the Center creates and fosters collaboration among industry, environmental organizations, academia, and government in each of its program areas and projects. The membership of the Center's Board of Trustees, its steering committees, and all its committees and working groups reflects its guiding philosophy: that all relevant parties must be involved if the complex issues surrounding environmental policymaking are to be resolved. The Center's mission is to identify emerging environmental issues, conduct related scientific research and economic analyses, and create and disseminate nonpartisan policy options for solving environmental problems.

About the Study on Economic, Environmental, and Social Outcomes of Dam Removal

The Heinz Center's study on Economic, Environmental, and Social Outcomes of Dam Removal was conducted under the terms of a joint project agreement between The Heinz Center, the Federal Emergency Management Agency, and the Electric Power Research Institute. This report does not necessarily reflect the policies or views of the study sponsors or of the organizations or agencies that employ the panel members.

Library of Congress Control Number: 2002103964

International Standard Book Number: 0-9717592-1-9

06 05 04 03 02 5 4 3 2

Printed in the United States of America

Additional copies of this report may be obtained free of charge from

The Heinz Center
1001 Pennsylvania Avenue, N.W., Suite 735 South, Washington, D.C. 20004
Telephone (202) 737-6307 Fax (202) 737-6410 e-mail info@heinzctr.org

This report is also available in full at www.heinzctr.org

Cover: Rindge Dam on Malibu Creek in California. Photo by Sarah Baish.

Heinz Center Staff

Thomas E. Lovejoy, President (from May 2002)
William Merrell, Senior Fellow and President (through April 2002)
Mary Hope Katsouros, Senior Fellow and Senior Vice President
Robert Friedman, Senior Fellow and Vice President for Research
Mary C. Eng, Senior Fellow and Treasurer
Jeannette Aspden, Fellow and Research Editor
Sarah Baish, Research Associate (through 1/15/02)
Melissa Brown, Staff Assistant
Kent Cavender-Bares, Fellow and Research Associate
Pierre-Marc Daggett, Research Associate and Travel Coordinator
Sheila David, Fellow and Project Manager
Judy Goss, Research Assistant
Cheryl A. Graham, Fellow
Daman Irby, Research Assistant
Jennifer Murphy, Research Associate
Robin O'Malley, Fellow and Project Manager
Jeffery Rank, Research Assistant
Elissette Rivera, Research Assistant
Carmen R. Thorndike, Executive Assistant

PANEL ON ECONOMIC, ENVIRONMENTAL, AND SOCIAL OUTCOMES OF DAM REMOVAL

William Graf (*Chair*), University of South Carolina, Columbia
John J. Boland, The Johns Hopkins University, Baltimore, Maryland
Douglas Dixon, Electric Power Research Institute, Gloucester Point, Virginia
Thomas C. Downs, Patton Boggs LLP, Washington, D.C.
Jack Kraeuter, Pennsylvania Department of Environmental Protection, Harrisburg, Pennsylvania
Mary Lou Soscia, U.S. Environmental Protection Agency, Portland, Oregon
David L. Wegner, Ecosystem Management International, Inc., Durango, Colorado
Philip B. Williams, Philip Williams & Associates, Corte Madera, California
Craig S. Wingo, Federal Emergency Management Agency, Washington, D.C.
Eugene P. Zeizel, Federal Emergency Management Agency, Washington, D.C.

Heinz Center Project Management Staff

Sheila D. David, Project Manager
Sarah K. Baish, Research Associate
Judy Goss, Research Assistant

Contents

Preface

Dams are the most common and widespread form of direct human control on river and stream processes. The construction, maintenance, operation, and potential removal of dams are critical aspects of scientific and policy discussions about rivers. Until recently, the installation of dams has been a widely supported method of river management in the United States. American rivers are collectively the most closely controlled hydrologic system of its size in the world. The nation now has the capability to store almost a full year's runoff in reservoirs behind more than 76,000 dams (counting those 6 feet high or more). Many of these structures have contributed to the economic development of the nation and the social welfare of its citizens. Irrigation water diverted from streams and temporarily stored by dams has supported agriculture in western states, and lock and dam structures sustain an inland water transportation system for bulk commodities worth billions of dollars throughout the nation. Dams can reduce flooding and provide water for consumptive uses (e.g., drinking) and non-consumptive uses (e.g., for power plants and other industrial cooling operations). Hydroelectric power from dams provides about 10 percent of the total electrical power for the nation, and in many locales, it is the primary source. The reservoirs created by dams provide recreational opportunities and prime waterfront property locations, with benefits enjoyed by millions of citizens. Small dams, often only a few feet in height, have been an integral part of the industrial, mining, agricultural, and urban history of the country.

The installation of dams and reservoirs to provide the economic and social services related to water has transformed the natural, interconnected river system of the United States into a partly artificial, partly natural regulated and segmented system. The environmental changes brought about by dams include drowning of channels and valued floodplains, with more than 600,000 miles of the nation's waterways under reservoir waters. Dams have changed downstream conditions, altering the physical bases of ecosystems in every region of the country. In concert with other human-imposed changes, especially those realized through river engineering and land use alterations, dams have contributed to the loss or change of riparian and aquatic habitat, including ecological systems that support endangered or threatened species of plants, animals, birds, and fishes. As these changes have become more apparent, many small and medium-sized dams have aged beyond their expected useful life spans, and for their physical safety must be repaired. Urbanization and other developments downstream from them have created hazardous conditions in some places. Changing economic conditions combined with aging and safety issues have made some dams obsolete, and new regulatory requirements cloud the future of others. Some dams are orphans, abandoned by owners who no longer have use for them. As a remedy for all these problems, the option of dam removal recently has become more widely considered.

After more than two centuries of policy attention almost exclusively to the building of dams, public decisions about removing some structures have drawn increasing interest because of the expense of maintaining antiquated structures. Philosophically, the United States has supported the intensive use of rivers for economic development throughout its history, but over the last few decades, growing concern about environmental quality, endangered species, and aesthetic characteristics of rivers has become more prominent in the national discourse. In many cases, these new emphases have become part of national, regional, and local policies. From a scientific perspective, recent research conducted by hydrologists, geomorphologists, and ecologists has begun to detail the changes brought about by dams. This knowledge is emerging in the early twenty-first century because many large dams did not begin appearing on the American landscape until about 1960. It has taken two or three decades for the physical and ecological consequences of the structures to become apparent.

If it is true that Americans now have considerable experience in building dams and assessing their effects, it is equally true that even the most expert have relatively little experience in removing dams and assessing the outcomes of their removal. While national attention has been focused on a few highly visible dam removal issues involving large structures, such as the dams on the Snake River in the Pacific Northwest, the removal of numerous small dams and a few medium-sized ones has continued apace. Although the precise number of dams removed from the nation's rivers is unknown, it certainly is at least five hundred. The number of candidates for removal is certain to increase as the structures continue to age, and as further emphasis on river restoration stimulates more interest in removal as one of a series of management options.

When dam owners, governmental agencies, interest groups, and private citizens debate removal options for specific structures, the decision-making process often needs to be reinvented for each case, with no accounting for scientific understanding of the likely outcomes of the decision. This report, which focuses on the removal of small dams (defined as storing 1–100 acre-feet of water), seeks to assist the decision-making process regarding dam removal by providing information for use by dam owners; policymakers; interest groups; private citizens; and personnel in local, state, and federal agencies. After providing extensive background and contextual information, the authors of this report strive to

- Outline the nature of likely environmental, social, and economic outcomes of dam removal
- Define indicators for measuring or monitoring environmental, social, and economic outcomes of dam removal
- Indicate sources of environmental, social, and economic data that may help place each specific case in context for decision makers

This report emphasizes the potential environmental, economic, and social science aspects of dam removal rather than the details of the decision-making process itself. The treatment of these scientific aspects is necessarily uneven because there is more direct scientific research available on the environmental dimensions of the issue, and relatively less about the economic and social dimensions.

The authors of this report were brought together by The H. John Heinz III Center for Science, Economics and the Environment as the Panel on Economic, Environmental, and Social Outcomes of Dam Removal. The panel included specialists in geography, economics, engineering, environmental law, state and federal administration, environmental consulting, hydraulic engineering, dam safety, hydropower, and aquatic ecosystem management. The panel met three times over the course of the 18-month study period, twice in Washington, D.C., and once in Southern California to visit field sites. The panel hosted several guests during its meetings to learn more about specific research activities related to dam removal and to receive the latest information about the subject. The Federal Emergency Management Agency, the Electric Power Research Institute, and The Heinz Center financially supported the activities of the panel.

The work creating this report was facilitated and coordinated by Sheila D. David, fellow and project manager for The Heinz Center. Her skillful planning, guidance, and management were critical to the successful completion of the project. She was a full and active partner along with panel members in the discussions and deliberations that went into the total effort. Sarah Baish, research associate for The Heinz Center, was a critical component of the project in managing the flow of ideas and paper, as well as writing case examples and making the essential arrangements for committee activities.

Individuals chosen for their expertise and diverse perspectives reviewed the report. Their independent review provided candid comments and suggestions that significantly improved the report. The panel wishes to thank the following individuals for their input during the review process: Syd Brown, California Department of Parks and Recreation; Charles C. Coutant, Oak Ridge National Laboratory; David Freyberg, Stanford University; Gordon E. Grant, U.S. Forest Service; Francis J. Magilligan, Dartmouth College; Larry Olmsted, Duke Power; A. Dan Tarlock, Chicago-Kent College of Law; and Chari Towne, Delaware Riverkeeper Network. Any errors or oversights in the final document are solely the responsibility of those who served on the panel.

This report does not advocate dam removal or retention in general or in any particular cases. There are numerous organizations and individuals who can speak to these viewpoints. Rather, this report is intended to be objective, and to offer the best science that is available in the belief that the best public policy decision is the one that is best informed.

William L. Graf
Chair

Acknowledgments

MANY INDIVIDUALS assisted the panel in its task by reviewing draft proposals for the project, recommending panel members, participating in panel meetings, providing data and background information to the panel, recommending individuals to be interviewed, or reviewing and editing drafts. The panel wishes to express its appreciation to the following people for their invaluable contributions to this project:

Jeannette Aspden, The Heinz Center, Washington, D.C.
Bruce Aylward, World Commission on Dams, Cape Town, South Africa
Mike Bahleda, Electric Power Research Institute, Blacksburg, Virginia
Donald Bathurst, Federal Emergency Management Agency
Angela Bednarek, University of Pennsylvania
Margaret Bowman, American Rivers, Washington, D.C.
Syd Brown, California Department of Parks and Recreation, Sacramento, California
Karen Bushaw-Newton, Academy of Natural Sciences, Philadelphia, Pennsylvania
Robert Friedman, The Heinz Center, Washington, D.C.
David Freyberg, Stanford University, Stanford, California
Suzanne Goode, California Department of Parks and Recreation, Calabasas, California
Robert Hamilton, U.S. Bureau of Reclamation, Denver, Colorado
Joan Harn, National Park Service, Washington, D.C.
Ed Henke, Historical Research, Ashland, Oregon
A. Paul Jenkin, Surfrider Foundation, Ventura, California
Reinard Knur, Anaheim, California
Elizabeth Maclin, American Rivers, Washington, D.C.
Laura Ost, Consulting Editor, Arlington, Virginia
Timothy Randle, U.S. Bureau of Reclamation, Denver, Colorado
Jason Shea, U.S. Army Corps of Engineers, L.A. District, Los Angeles, California

Summary

When one tugs at a single thing in nature, he finds
it attached to the rest of the world.

—*John Muir*

Dams are common features of the American landscape and waterscape, forming an integral part of the nation's infrastructure that contributes to the collective economic and social welfare. The construction and operation of dams also have imposed environmental, economic, and social costs that only recently have become clear. Interest in dam removal is a recent outcome of the aging of many of the structures, evolving societal values, and increasing scientific knowledge about changes brought about by dams.

Throughout its history, the United States has supported the intensive use and development of rivers for economic gain. Americans traditionally have viewed rivers as water resource-related commodities to be used rather than as ecosystems to be protected. However, in the past few decades, growing concern over environmental quality, endangered species, and aesthetics of landscapes has become more prominent in the national discourse about rivers. Also evident are concerns about dam safety and security, downstream risks related to unsafe dams, and the future of structures that have become obsolete. The environmental and safety issues associated with dams have become components of local, regional, and national policies.

The majority of dams in the United States are small, storing less than 100 acre-feet of water. Private individuals, firms, or local entities own most of these small structures, although some orphan dams lack any formal, established ownership. An unknown number of dams already have been removed, likely more than 500 mostly small, run-of-river structures. Many of these removals were the products of decisions by individual owners who sought a variety of economic benefits, although the environmental reasons for dam removal are numerous and often supported by local or state governments. The decision to remove a dam by its owner may not be made in the public arena. However, because of state and federal regulations, the decision to approve a removal becomes a public process.

The Heinz Center Panel on Economic, Environmental, and Social Outcomes of Dam Removal generated this report to assist dam owners, private citizens, and other decision makers. It outlines the current state of research and knowledge related to dam removal and recommends steps and indicators for decision making regarding dam removal. This report is a primer, designed to provide background information and basic principles derived from science and experience for decision-making processes. For the purposes of this report, the panel defined the following dam size categories based on reservoir storage rather than height or other measures because the size of the reservoir is related most directly to the magnitude of potential effects on river hydrology:

Small: reservoir storage of 1–100 acre-feet

Medium: reservoir storage of 100–10,000 acre-feet

Large: reservoir storage of 10,000–1,000,000 acre-feet

Very large: reservoir storage of greater than 1,000,000 acre-feet

This report focuses on small dams, because historically Americans have the most experience with the removal of such structures, and this size of dam is most likely to be considered for removal at present. The report addresses medium-sized structures but in less detail, because only a few are under consideration for removal. Lessons learned from the removal of small structures may provide useful input, with some modification, for decisions about larger dams, as owners/operators express interest in their removal. The report does not address the potential removal of large or very large multipurpose dams. The issue of removing large dams, such as the Snake River dams, is being considered in detail by the U.S. Army Corps of Engineers (2001) through an environmental impact statement process. A similar process exists under the Federal Energy Regulatory Commission for private hydropower dams. Within this context, the panel sought to address a three-part task: (1) outline the nature of likely environmental, economic, and social outcomes of dam removal; (2) define indicators for measuring and monitoring outcomes; and (3) indicate sources of useful information for researchers and decisionmakers considering dam removal. The Heinz Center panel was charged with investigating the outcomes of dam removal and did not evaluate alternatives to removal. These potential alternatives include re-engineering the dam structures, changing operating rules, constructing fish passages, sediment management, and conducting other mitigation measures focused on habitat.

The nation has many small dams that are abandoned or obsolete and whose owners may wish to consider removal as a viable option. Neither the panel nor this report advocates any particular position regarding the advisability of removal or retention of dams. The panel seeks to help resolve potential conflicts that are likely to develop in balancing societal and environmental needs with respect to dams. The report does not make recommendations about individual structures. Rather, this report recounts the lessons learned in previous dam removals and scientific investigations as an aid to informed, reasonable decision making. The panel believes that dam owners, private citizens, researchers, and other decision makers are more likely to reach conclusions that serve the best interests of all community members if they have the best available methods and information. The panel offers this report as a contribution to achieving the goal of informed, effective decision-making processes.

BACKGROUND

The National Inventory of Dams,* a database maintained by the Corps of Engineers and Federal Emergency Management Agency, catalogs more than 76,000 dams in the United States that are 6 feet high or more and impound at least 50 acre-feet of water, are 25 feet high and impound at least 15 acre-feet, or pose a serious hazard to people downstream. The potential storage behind these dams is almost equal to the nation's total annual runoff. About one-quarter of all dams were constructed during the

* The inventory is available online but the site was taken offline as a security precaution after the September 11, 2001, terrorist attacks. The site may be restored after further evaluation. The Web site is http://crunch.tec.army.mil/nid/webpages/nid.cfm.

1960s, and many structures now are half a century old. Reasons for building these dams included

- Water supply for domestic and industrial use
- Irrigation water supply for agriculture
- Flood suppression
- Waterpower (mills)
- Hydroelectric power
- Navigation
- Flat-water recreation
- Waste disposal

There is no completely accurate accounting of the number of dams removed in the United States, because accurate records of historical removals are rare. American Rivers Incorporated has documented the removal of almost 500 structures, though the actual total is likely to be much larger. Almost all dams removed so far have been small, privately owned ones that are most often of the run-of-river type, although a few medium-sized dams with some storage also have been removed. Reasons for dam removal include

- Economic obsolescence
- Structural obsolescence
- Safety considerations
- Legal and financial liability
- Dam site restoration
- Ecosystem and watershed restoration
- Restoration of habitat for riparian or aquatic species
- Unregulated flow recreation
- Water quality or quantity

DAM REMOVAL DECISIONS

A key premise of this study is that better decisions will be made about whether to retain or remove a dam if the process is logical, defensible, and organized. The decision to remove a dam by its owner may not be made in the public arena. However, because of state and federal regulations, the decision to approve a removal becomes a public process. Such a process would begin with the owner's desire to remove a dam. The next step would be the identification of the specific goals that the owner and/or the communities involved with the dam hope to achieve. Public discussions about the advantages and disadvantages of retention versus removal are required, with freely available information, often assembled in map-based formats. Reliable maps and data about many of the environmental, social, and economic aspects of decisions related to dam removal are available from the World Wide Web (site addresses are given in Appendix A of this report).

The panel designed and advocates a systematic approach to decisions about dam retention or removal (Figure S.1). The steps include the following:

1. Establish the goals, objectives, and a basis for the decision, a task that includes the collection of information about the environmental, social, economic, regulatory, and policy contexts for the decision and its outcome.
2. Identify major issues of concern, ranging from the safety and security of a dam to those related to the cultural interests of the population involved.
3. Assess potential outcomes and gather data about the operations of the river; the dam; the legal regime; and the ecological, social, and economic systems associated with these elements. These assessments depend on the evaluation of a series of indicators that provide insight into present and likely future conditions.
4. Make decisions within a framework that encompasses available knowledge about the gains and losses, costs and benefits, public support and concerns, and private and public interests.

A key component of this step-by-step process is the gathering of data and assessment of outcomes, which not only provides a view of the present conditions, but that also may be useful in describing the likely future conditions once the dam is removed. Decision makers can use this information to assess the "with dam" and "without dam" future scenarios and consider what might happen in the short term (a few months), medium term (a few years), and long term (a few decades). The panel developed an extensive list of issues and associated indicators that can be measured in the present and predicted for the future (Box S.1). See Table 3.1 on pages 90–93 for an extended list of indicators.

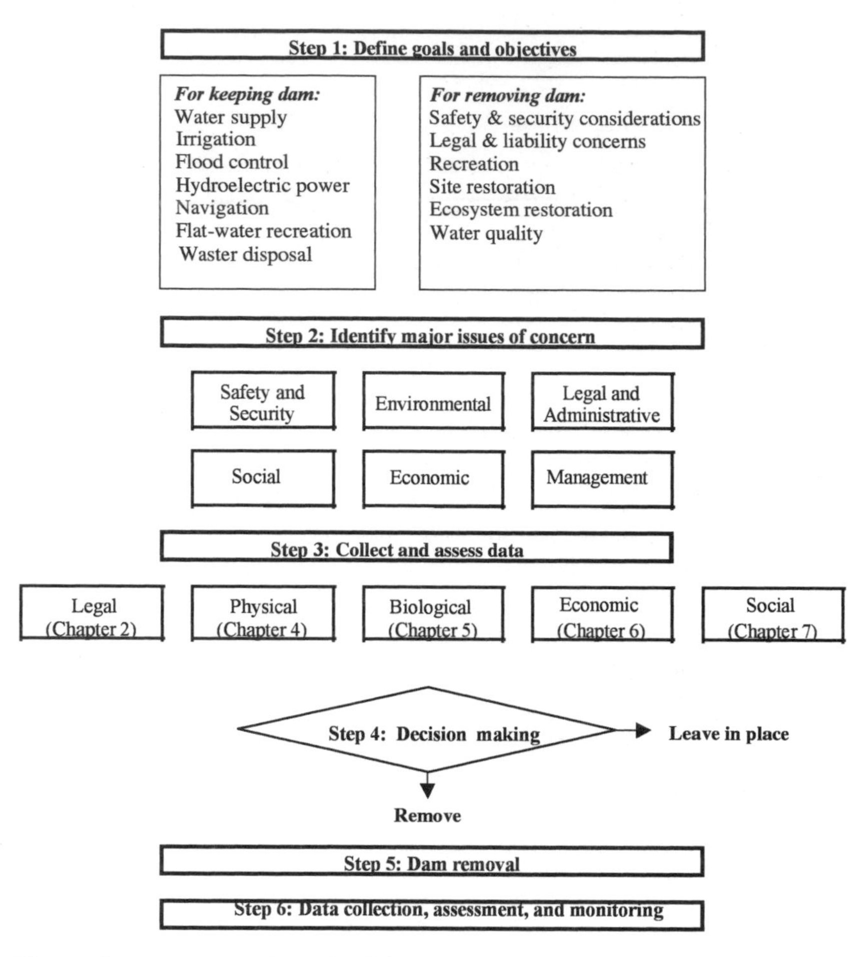

Figure S.1 A general method for dam removal decisions

PHYSICAL ENVIRONMENTAL OUTCOMES OF DAM REMOVAL

Dam removal can restore some but not all of the characteristics of the river that existed before the dam was built. Dam removal creates a more natural river than existed with the dam in place because some aspects of physical integrity* are restored to the river downstream from the dam site.

* The word integrity is especially apt when applied to rivers because it means unity, completeness, and the quality of state of being complete or undivided (Websters New Collegiate Dictionary, 1981).

Box S.1 Key Indicators for Dam Removal Decisions[a]

Physical
- River network segmentation
- Watershed fragmentation
- Downstream hydrology
- Downstream sediment system
- Downstream channel geomorphology
- Floodplain geomorphology
- Reservoir geomorphology
- Upstream geomorphology

Chemical
- Water quality
- Sediment quality (reservoir area and downstream)
- Air quality

Ecological
- Aquatic ecosystems
- Riparian ecosystems
- Fishes
- Birds
- Terrestrial animals

Economic
- Dam-Site economics
- Economic values, river reach
- Regional economic values

Social
- Safety and security
- Aesthetic and cultural values
- Non-majority considerations

[a] Ideally, these indicators would be used to measure or estimate today's conditions and forecast conditions one year, five years, and a decade or two into the future.

In addition to the effects of their reservoirs, which inundate terrain and ecosystems, dams affect physical integrity by fragmenting the lengths of rivers, changing their hydrologic characteristics (especially peak flows), and altering their sediment regimes by trapping most of the sediment entering the reservoirs. These effects cause downstream landscape changes, including channel shrinkage and deactivation of floodplains.

Dams also cause water quality changes that alter aquatic ecosystems. The removal of dams has the effect of reversing some undesirable changes, subject to the limits imposed by many other human influences on the watercourse. The most important positive outcome of dam removal is the reconnection of river reaches so that they can operate as an integrated system, which is the basis of a river with restored physical integrity. Productive, useful ecosystems can result from dam removal, but predictions of outcomes are sometimes difficult because of the many interrelated changes in physical and biological systems caused by placement of the dam and other physical stresses on the river. For example, dam removal may result in the remobilization of contaminated sediments once stored in reservoirs.

BIOLOGICAL OUTCOMES OF DAM REMOVAL

One way to learn about the potential effects of dam removal is to review what is known about the effects of dam installation on a river system. Although the changes brought about by installation may not be completely reversible, they do help to predict the various consequences of removal.

Changes in the physical system of a river imposed by a dam, and partly reversed by dam removal, cause associated adjustments in the biological components of the ecosystem. These biological changes, particularly among fish and macro-invertebrates, include altered movement patterns, residence times, and general habitat opportunities. These biological ecosystem changes are variable in time and space. The extent and intensity of the changes depend on the size of the dam (storage capacity), quantity and quality of sediment in the reservoir, timing of reservoir level fluctuations, limnological conditions in the reservoir, and stability of the downstream river reach. Non-native exotic species also affect native species in both rivers and reservoirs. Dam removal may, in some cases, increase the abundance and diversity of aquatic insect, fish, and other populations, but long-term data and numerous "before and after" tests of population trends are not available. Reservoirs create wetland areas in some cases; the removal of a dam and draining of a reservoir may create some wetlands downstream, but at the expense of some wetlands upstream. Dam removal often results in the replacement of one aquatic community with another that is, therefore, partly natural and partly artificial. The most significant biological effect of the removal of small structures is the increased accessibility of upstream habitat and spawning areas for migratory and anadromous fishes.

ECONOMIC ASPECTS OF DAM REMOVAL

From an economic standpoint, dam removal is not unambiguously good. Economic analysis can be helpful for setting priorities and facilitating communication among stakeholders and agencies. Benefit–cost analysis provides a process for identifying and measuring the outcomes of dam removal, whether they are perceived as positive or negative, and for clarifying trade-offs in the decision-making process. Traditional benefit–cost approaches are imperfect for dam removal, however, for several reasons. In traditional analyses, there is a "no action" alternative, which serves as an economic baseline that is the starting point for measuring beneficial and adverse effects. In many dam removal decisions, there is no such baseline, because "no action " (i.e., no project) is not possible. The owner of a dam may be compelled by safety or economic considerations to either remove the dam or repair it, and therefore a nontraditional reference case is required. Additionally, many environmental outcomes are uncertain or difficult to establish in monetary terms. Even so, they had best not be ignored, because they are among the primary concerns in public discourse and debate about dam retention or removal. Reasonable valuations of outcomes that are rooted as firmly as possible in economic theory and applications offer the best path to economically informed decisions.

SOCIAL ASPECTS OF DAM REMOVAL

Little research has been conducted to date on the social science aspects of dam removal. This is a serious shortcoming, because the social context of dam removal decisions is often as important as the environmental and economic contexts. Social outcomes of dam removal include, for example, the aesthetics of the dam site and adjacent river reaches. There may be a clash of values; some stakeholders may emphasize their desire for a partially restored environment, whereas others may warn against the loss of a historically significant structure or water body. On the other hand, the draining of a reservoir may restore a historical landscape. Cultural values associated with human and natural landscape components are likely to be important in discussions related to potential dam removals. Water rights, property values, tribal rights, and the maintenance of storage capability are also likely to be issues, along with improved water quality and changed recreational opportunities.

CONCLUSIONS AND RECOMMENDATIONS

The Heinz Center panel identified conclusions and recommendations in three general categories: making decisions today, data needs, and improving tomorrow's decision making.

MAKING DECISIONS TODAY

Dam removal decisions require careful planning and review. To be effective and useful for managers, decision makers, and the public, a removal project needs to be scientifically based. Decisions about dam removal also take place in specific economic and social contexts that need to be taken into account. Decision-making processes for dam removal are, in most cases, more effective when they are systematic, open, and inclusive of the people in the affected communities.

- **The panel recommends that participants in public decision-making processes use a multistep process similar to the one outlined in this report (Figure S.1), beginning with the establishment of goals as a basis for the process, and including the identification of the full range of interests and concerns of those likely to be involved, assessment of potential outcomes, and informed and open decision making.**

The assessment of potential outcomes of dam retention or removal requires measurable indicators that can be used to assess the present environmental, economic, and social conditions associated with the dam and to monitor future changes.

- **The panel recommends that assessment of potential outcomes of a decision to retain or remove a dam include the evaluation of as many indicators as are applicable to the situation, with the assessment conducted for short-, medium-, and long-term periods,**

and for the "with dam" as well as "without dam" alternatives. The panel developed a list of measurable indicators (Box S.1 and Table 3.1) that can be used to support the decision-making process outlined in Figure S.1.

Decisions to remove dams in a complicated physical and biological system can have far-reaching implications both upstream and downstream. The consideration of a limited scope of outcomes is likely to have unforeseen consequences.

- **The panel recommends that a dam removal decision take into account watershed and ecosystem perspectives as well as river-reach perspectives and the more limited focus on the dam site.**

Data Needs

Data on dams that have been removed can be useful to decision makers considering the fate of existing structures, yet there is no centralized mechanism for collecting, archiving, and making available such information on a continually updated basis. The effects and effectiveness of any individual dam removal depend, in part, on the nature of the rest of the affected river system. There is an obvious need for a geospatial database that provides accurate, readily accessible data on the segmentation of the nation's river systems by dams and the quantity and quality of sediment discharged in the nation's rivers. In addition, monitoring after dam removal is essential to enable stakeholders to evaluate whether the goals and objectives of the removal have been met.

- **When dams are removed, their entries in the National Inventory of Dams are deleted and the National Performance of Dams Program retains information about them. The panel recommends that federal agencies improve the availability of information about dam removal by making this database widely known and available to the public.**

- **The panel recommends that the U.S. Geological Survey maintain and extend its network of sediment measurement statistics throughout the total national stream gauging system.**

- **The panel recommends that the U.S. Environmental Protection Agency and/or U.S. Geological Survey consider augmenting the existing national stream-reach geographical data to include the location of dams to allow better analysis and understanding of the segmented nature of the nation's streams and rivers.**

- **The panel recommends that the U.S. EPA and/or appropriate state and local governmental agencies conduct a monitoring and evaluation program following dam removal. This program should be developed and implemented so that vital data on the natural and enhanced restoration of habitats is collected and made available in public datasets for use in adaptive management.**

Improving Tomorrow's Decision Making

Dams are a ubiquitous feature of the American landscape and waterscape and form an integral part of the nation's economic infrastructure. The building of these structures has produced significant economic benefits, but the effort also has imposed environmental, economic, and social changes and costs. Science to support decisions about dam removal is progressing, but there is little cross-disciplinary communication, and research priorities have not been established to guide researchers or funding efforts.

- **The panel recommends that federal agencies and other organizations consider sponsoring a conference for researchers who currently focus on the scientific aspects of dam removal with the specific objectives of improving communication across disciplinary boundaries, identifying gaps in knowledge, and prioritizing research needs. The conference should not be a forum for debating whether dams should be removed but rather should focus on science and the state of knowledge available for decision makers.**

Several fundamental technical aspects of dam removal are poorly understood. Dam removal may result in the remobilization of contaminated sediments once stored in reservoirs, yet understanding of sediment processes is poor. Sediment quality and quantity are the most important issues in considering biophysical outcomes of dam removal. Other issues include vegetation changes, bank erosion, channel change, and effects on groundwater. Water quality is an important human health and environmental concern, yet outcomes of dam removal on water quality are poorly understood. One of the most important outcomes of dam removal is the reconnection of river reaches so that they operate as a free-flowing, unobstructed system—that is, restoring the physical integrity of the river system. However, empirical data are lacking on river channel changes downstream from removed structures.

■ **The panel recommends that the scientific community of river researchers provide (1) improved understanding of sediment quality and dynamics to provide a scientific basis for evaluating contaminated sediments, (2) improved understanding of the roles that dams and their potential removal play in water quality, (3) empirically derived explanations of river channel change upstream and downstream from removed dams, and (4) a knowledge base of the likely fate of sediments and their contaminants downstream from removed dams.**

Formal economic analyses can be very helpful in supporting the decision-making process for dam removal, in setting priorities, and, most of all, in facilitating communication among stakeholders and agencies. Nevertheless, significant challenges remain for those who would use methods such as benefit-cost analysis for this purpose. Dam removal has various environmental effects, including some that are highly uncertain and difficult to quantify. It may be tempting to ignore these issues, as often was done in the earlier building of dams. However, these non-quantified environmental effects are major issues for consideration when dealing with a possible dam removal and had best not be ignored. The science of economics does not yet offer decisionmakers considering dam removal a sufficient array of analytic tools and supporting data to assess adequately the economic outcomes of a decision in quantitative terms.

■ **The panel recommends that the community of economics researchers provide (1) improved economics evaluation tools to enable the assignment of monetary valuations for outcomes of dam removal, and (2) empirical research on changes in property values associated with dam removals already accomplished.**

The social outcomes of dam removal decisions are not yet well known, but standard social science, survey-based research can help stakeholders understand potential changes in individual and community behavior related to such decisions. The adaptive management process for environmental systems could be extended to social systems so that river managers would be able to make informed adjustments to their plans.

- **The panel recommends that agencies and organizations that fund social science research support investigations into the social and cultural dimensions of cases in which dams already have been removed, as a way of improving the predictability of outcomes.**

- **The panel recommends that decision makers in dam removal cases should undertake social impact studies modeled on the environmental impact studies that are a common feature of such decision-making processes. These social impact studies should address the cultural significance of the dam site (e.g., as a tribal sacred site), reservoir area, and river areas likely to be changed by the proposed removal.**

Dams are important parts of the nation's economic and historical fabric, and their presence affects everyone's lives. Dams are also integral parts of the nation's riverine ecosystem, exerting wide-ranging changes in the physical and biological processes in rivers. A decision to remove or retain a dam has implications for a variety of community and national values, some of which may not be complementary. The surest route to a successful, informed decision is to explore the likely environmental, economic, and social outcomes before the decision to retain or remove a dam is made.

As a follow-up on the activities related to this project, the Heinz Center proposes to host a conference for researchers on the science of dam removal with the objectives of clarifying the present state of knowledge in the various scientific disciplines addressing the issue, identifying topical areas in which one discipline can assist another in problem solving, and specifying the gaps in knowledge that require additional research to better support decision making. The Center also seeks to apply the concepts and procedures outlined in this report to several test cases in which dam removals are being considered. The Center also sees the need for a study and report that provides alternatives to dam removal, to aid owners of small dams and public decision makers, especially with cases of abandoned or orphaned dams.

1

Introduction and Background

Dams are the fulcrums of many of the increasingly important environmental quality decisions facing the nation's river managers and the public. The estimated 76,000 dams (counting those 6 feet high or more) constructed in the United States have transformed the nation's rivers and greatly influenced the economic development and social welfare of its citizens. Over the last 200 years, dams have been built and operated for a variety of purposes: to reduce flood flows, provide agricultural and urban water supply, control fires, improve navigation, offer recreational opportunities, and generate electricity. Dams also have created new habitats, such as nesting areas for riparian birds and migratory waterfowl on reservoir deltas, and lake fish habitat. However, some dams have created long-lasting changes in the quality of riparian and aquatic habitats and have contributed directly to the decline of some commercially important fish as well as endangered species. Increasingly, river management debates include discussions about dams and, in many cases, their removal or alteration.

The nation now has the capability to store the equivalent of almost one full year's runoff in reservoirs behind more than 80,000 structures. If the definition of "dam" includes the smallest structures, there may be as many as 2 million (Graf, 1993). At first glance, the long-term costs and benefits of dams seem straightforward, but they are actually difficult to determine. Dams have not completely controlled floods, but some have significantly reduced loss of life and provided property protection. Irrigation waters diverted from streams and temporarily stored by dams have stimulated the agricultural and economic development of western states. Lock and dam structures in the Mississippi basin have created an inland water transportation system for bulk commodities worth billions of dollars. However, this system has relied on the continued maintenance of federally funded river engineering works. Dams generate more than 10 percent of the nation's electricity and more than 70 percent of the electricity in the Pacific Northwest.

Many U.S. dams were constructed during the late nineteenth century and early to mid-twentieth century. Dam building accelerated in the early 1930s, and, by 1945, Grand Coulee Dam and Hoover Dam were the two largest power sources in the world (Costenbader, 1998). Hydroelectric dams provide electricity that generates power with far fewer air emissions (little or no carbon dioxide) or solid and liquid wastes than do most other sources of energy. The installation of dams and reservoirs to provide electricity, recreation, and property protection from floods has transformed the natural, interconnected river system of the United States into a fragmented—and partly artificial, partly natural—system of river segments. The environmental changes associated with dams include the loss of channels and associated floodplains, with more than 600,000 miles of the nation's rivers under reservoir waters (Huntington and Echeverria 1991). Dams have resulted in changes in biology and biological processes, and they have altered the hydrologic and physical bases of ecosystems in every region of the nation. Dams are features of the landscape everywhere, with the greatest density of dams in the eastern and southeastern states, and the greatest influence on the hydrologic system in the interior areas of the West (where structures store almost four years' runoff).

The recent attention to the effects of dams stems from changing social values, dam safety issues associated with aging structures, and a general increase in the knowledge and scientific base of understanding of the long-term physical and ecosystem response. The nation previously supported the intensive use of rivers for economic development. In the last three decades, however, growing concern over environmental quality, mounting flood losses, endangered species, and aesthetic characteristics of landscapes have become more prominent in the national discourse about rivers. It also has taken two to three decades for some of the environmental changes caused by the larger structures, many built after 1960, to become apparent.

This report addresses downstream restoration and other changes that follow a dam removal. Restoration of the former reservoir is, of course, also a consideration. Frequently, the length of river involved in the reservoir area is short compared to the affected downstream areas. A relatively short reach of river upstream from the reservoir site is likely to be affected first by the filling of the reservoir and then by its draining. Accumulated sediments in the reservoir area and immediate upstream reach may be eroded and removed with the dam, though some remaining sediments may become the site of a new channel and near-channel landforms.

The downstream alteration by dams of the physical operation of rivers has resulted in changes in river landscapes, loss of riparian and aquatic habitat, fragmentation of migration corridors (especially for salmon and shad), and endangerment of threatened native fishes and riparian birds. The recovery of these endangered species may depend on removing or re-engineering dams or changing their operating rules, measures that bring about unavoidable conflict with the objectives for which the dams originally were built.

Federal environmental legislation relevant to dam operations and removal include the Endangered Species Act of 1973 (P.L. 93-205), Clean Water Act (originally called the Federal Water Pollution Control Act [*United States Code*, Title 33, Section 1251 et seq.*] and amended a number of times), National Environmental Policy Act of 1969 (P.L. 91-190), Wild and Scenic Rivers Act of 1968 (P.L. 90-542), and tribal laws. These and other relevant laws are discussed in Chapter 2. The recovery of riverine endangered species and commercial fisheries may hinge on some actions involving dams, and the Clean Water Act stipulates that it is national policy to restore and maintain the biological, chemical, and physical integrity of rivers, a task that also engages dams. Actions involving dams are usually limited to the removal of structures, but they may be extensive in their effects. The removal of a single small dam in a key location may free many miles of newly accessible spawning reaches. For example, the removal of the 7-foot-high Quaker Neck Dam on North Carolina's Neuse River system opened 1,000 miles of upstream spawning reaches for migratory fish, and the removal of Columbia Falls Dam opened access to 28 miles of Maine's Pleasant River. Although the present debate seems to pit social and economic benefits against these types of environmental goals, it is likely that some dams can be operated to benefit both socioeconomic and environmental ends.

Very large dams (dam size categories are defined in the next section) are generally not targeted for removal and are largely owned by the federal government. Companies or cooperatives privately own most

* Henceforth, references to the *Code* will be abbreviated using the format 33 USC §1251.

medium sized dams used for irrigation, water supply, hydroelectric power, and direct hydropower (e.g., for mills). A small percentage of medium sized structures are nonfederal hydropower dams licensed by the Federal Energy Regulatory Commission (FERC) and are periodically considered for relicensing.

Almost all small dams are privately owned, although some are owned by state, federal, or local agencies. Some small dams are orphaned (or abandoned) and may be taken over eventually by the state or local community. Structures of this size were constructed primarily for water diversion and irrigation purposes, to generate locally marketed hydroelectric power, to improve navigation on small and medium-sized streams, or to power machinery directly. Other small dams were constructed for recreational purposes. Many of these structures are in poor condition and no longer perform their original functions because of the efficiency of competing regional power grids, changing transportation needs that eliminated water transport on small and medium sized streams, and the economic decline of water-powered industries. Private owners may seek to remove dams because of safety concerns, high insurance costs, and maintenance costs. The potential removal of small structures can be a key step in river and riparian restoration, improved recreational opportunities, increased access to spawning grounds for anadromous fishes, and resolution of safety issues. Privately owned off-stream tailings dams that impound mining waste pose special policy challenges.

Regardless of size, all dams encounter safety issues deriving from the 1972 National Inventory of Dams Act (P.L. 92-367), which requires periodic inspections of all dams in the country. State inspectors evaluate each dam to assess the potential for loss of life and damage to property should the dam fail or be operated improperly. Their reports to the U.S. Army Corps of Engineers (USACE) and Federal Emergency Management Agency (FEMA) show that 14 percent of all dams in the country are rated as "high hazard" (indicating the potential loss-of-life hazard to the downstream area resulting from failure or misoperation of the dam), with an additional 18 percent rated as "significant hazard." Concerns about dam safety are related to the structures, but if a dam is removed, new river safety and flood hazard issues need to become part of the decision making process. Dams are the most common and widespread direct human control on river processes in the United States, and as such, their management, operation, construction, maintenance, and potential removal are all critical aspects of any scientific or policy debate about the future of rivers.

PURPOSE AND SCOPE OF THE HEINZ CENTER STUDY

In the 1960s and 1970s, pioneering, multi-objective research was undertaken to ensure the economic efficiency and productivity of proposed dams. Despite this effort, relatively little work is available to guide decision makers who seek a balance among the social, economic, and environmental consequences of dam removal. Part of the problem with current discussions about dam removal is the lack of formal frameworks for such evaluations, the lack of general agreement on useful indicators or data, uncertainty with regard to the environmental benefits to be gained or lost, and limited knowledge of available alternatives. It is possible to measure the economic productivity derived from dams, particularly in terms of water delivery, hydroelectric power, recreation, and navigation. Non-use values, and values for wildlife and restored, more natural landscapes, are more elusive and difficult to quantify.

Discussions with experts on river restoration, hydropower, water supply, dam removal, and dam safety led the Heinz Center staff to believe that a review and study of potential outcomes, guidance to useful sources of information, and insights into current knowledge regarding dam removal would assist decision makers and help them to make more informed decisions. The Panel on Economic, Environmental, and Social Outcomes of Dam Removal was convened to conduct this study. *Neither the panel nor this report advocates any particular position regarding the advisability of removal or retention of dams. The report does not recommend decisions that should be made about dams collectively throughout the nation or about individual structures.* Rather, this report is intended to aid informed, reasonable decision making by recounting the lessons learned in previous dam removals and scientific investigations. The panel offers this report as a primer, a contribution to achieving the goal of informed, effective decision-making processes. This report builds the necessary informational foundation for researchers and decision makers by focusing on the following objectives:

1. *Outline the wide-ranging outcomes of dam removal, including potentially positive and negative effects, and a list of issues to be addressed in the decision-making process.* Examples of outcomes include the upstream and downstream geomorphic, hydrologic, and biological effects; changes in the economic infrastructure at the local level; and elimination of established recreational oppor-

tunities along with creation of new but different opportunities. The list of outcomes will be as specific and complete as possible, but it is unlikely that all effects will be important for every dam.

2. *Define indicators for measuring and monitoring environmental, economic, and social factors related to dam management and/or removal.* Examples include environmental indicators such as stream flow, water quality, sediment loads, and species diversity and abundance for aquatic and riparian terrestrial ecosystems. Economic indicators may include employment data, transportation planning issues, investment opportunities, and land parcel valuations. Social indicators might include recreational opportunities, population distribution, and quality of life measures. Indicators will be those most readily available and most easily measured; they will be informative for experts and understandable to educated laypersons.
3. *Provide available information sources for decision makers.* Information available to support decisions regarding whether or not to remove a dam is scattered among a variety of public agencies and private, nongovernmental organizations. This report provides a list of information sources and ongoing scientific research related to dam removal, and, if available, data sources such as World Wide Web sites and/or the names and locations of researchers and the topics of their research.

This report focuses on small dams because these structures are of most widespread interest now for possible removal. The size of dams can be defined in a number of ways, such as by height or width, but the most useful definition is reservoir storage capacity. The capability of a dam to store water (and, inadvertently, sediment) is a rough measure of its potential hydrologic impact. For the purposes of this study, dams are characterized as follows:

Small: reservoir storage of 1–100 acre-feet

Medium: reservoir storage of 100–10,000 acre-feet

Large: reservoir storage of 10,000–1,000,000 acre-feet

Very large: reservoir storage of more than 1,000,000 acre-feet

Another reason for focusing on small dams is that almost all dams removed so far have been small, and, therefore, almost all the present opportunities to evaluate the effects of dam removal scientifically are limited to this size range. Although the majority of dams under consideration for removal are small, some medium-sized structures are under active consideration as well; Matilija Dam in California is being considered for dismantlement, and some others, such as Condit Dam in Washington, are in the advanced planning stages for removal. Lessons learned from the removal of small structures may be useful in the future, if more medium-sized structures are considered. Only two large dams are currently under active consideration for removal: Englebright Dam on the Yuba River in California and Glines Canyon Dam on the Elwha River of Washington. Additional large dams on the Snake River in the Pacific Northwest may be reconsidered for removal after a multi-year period for mitigation tests. The National Marine Fisheries Service and U.S. Fish and Wildlife Service will monitor fish runs and, if no improvement is seen, reconsider dam removal.

This report is aimed at decision makers and policymakers, dam owners, and planners at the federal, state, and local levels who are interested in learning how to make decisions that take into account the economic, environmental, and social aspects of dam removal in the United States. The audience includes legislators who establish broad policy and programs and local government officials who develop and implement policies regarding land use, endangered species, dam safety, and water power and supply. Citizens concerned about dam removal, and social and natural scientists, will find this report informative and helpful in determining new research needs.

The momentum of dam removal discussions is increasing, and other organizations were studying this subject at the same time as the Heinz Center. For example, American Rivers, Friends of the Earth, and Trout Unlimited (1999) issued a cooperative report outlining the experience of specific dam removal projects. The report is available as a paper-covered book and is on the Web at http://www.americanrivers.org/damremovaltoolkit/successstoriesreport.htm. The Aspen Institute began a dialogue on dams and rivers in September 2000; about 30 people have been convening every few months to consider and recommend guidelines for decisions regarding dam removal. The Aspen dialogue was expected to end by September 2002. In addition, the World Commission on Dams (2000) recently issued a major report on decision making regarding dams and economic development. The report is available as a paper-covered book and in digital form from the commission Web site: http://www.dams.org/report.

CENSUS OF DAMS IN THE UNITED STATES

Scientific research related to dam removal and the supporting decision-making process takes place in a historical and geographical context. As noted earlier, many U.S. dams were constructed during the late nineteenth or early to mid-twentieth century, and their presence has become commonplace. Rural and urban dwellers have come to rely on reservoirs for a constant supply of water and electricity. Because of the perceived permanence of dams, many people believe that existing dams will remain unchanged, despite the limited life expectancy of many small and medium-sized structures due to aging and reservoir sedimentation (some conceivably could last much longer if properly maintained). Information about the general background of dams that form the national infrastructure, and the reasons for past decisions that resulted in the present arrangement of dams, can be helpful to those seeking to understand the environmental, economic, and social implications.

The total number of dams that have been built on the rivers of the United States is unknown. Accurate records, especially for small structures, are lacking, and a national accounting would be an enormous undertaking in data collection and management. The best available data are in the National Inventory of Dams (NID), developed from the first broad-based effort to collate information about dams on a national basis as defined by the National Inventory of Dams Safety Act (P.L. 92-367) and signed into law by President Nixon in 1972. The collapse of Teton Dam on Idaho's Teton River in 1976, and the attendant loss of life and property, stimulated further interest in cataloging the nation's dams as potential hazards. In 1986, additional legislative emphasis on building a database appeared in the Water Resources Development Act (P.L. 99-662). The National Dam Safety Program Act of 1996 (P.L.104-303) supports states in their regulation of dams. The result of these pieces of legislation was the NID, which is managed by the USACE and coordinated by the FEMA. These agencies published early digital versions of the database on CD-ROM (Federal Emergency Management Agency and U.S. Army Corps of Engineers 1994, 1996); more recent versions of the inventory are usually available on the World Wide Web. The following discussion is based largely on the 1996 CD-ROM version (subsequent revisions have been relatively minor, though there are continuing additions to the data, mostly for small dams).

The enabling legislation defined the dams eligible for inclusion in the NID as those structures whose collapse might pose a threat to life and property downstream, those greater than 6 feet high with more than 50 acre-feet (61,000 cubic meters) of storage, and those that are 25 feet high with more than 15 acre-feet (18,500 cubic meters) of storage. In 1996, approximately 76,000 dams were included in the NID; that total has grown slowly since then, as more data have been made available by states. In addition to the structures included in the database, there are numerous small dams on the nation's small watercourses. A report by the National Research Council (1992) states that there are well over 2.5 million dams in the United States.

Sizes of Dams

From an engineering perspective, a most informative way of measuring the sizes of dams is to describe the physical dimensions: height, width, and thickness, for example. When considering dam removal, however, the storage volume behind the structure is a more useful measure of size because it is a direct measure of the hydrologic and sedimentary effects of the dam. The larger the storage volume, the greater the downstream effect of the structure on sediment throughput.

Many small dams have little or no storage, are informally designed, and age poorly. Medium-sized dams are often single-purpose structures erected with considerable investment, whereas large dams are multipurpose, large-scale engineering projects of regional or national significance. Taken together, the 76,000 dams in the NID have a storage capacity that is nearly equal to the nation's mean annual runoff, but the distribution of this storage volume among the various sizes is unequal (Graf, 1999). The majority of dams in the United States are in the small size range, but they store very little water and sediment. From a national perspective, the greatest proportion of the total volume of reservoir water is stored behind the large dams (Figure 1.1).

Types of Dams

From the standpoint of function, there are two general classes of dams: those that are designed to store water and those that are not. Storage*

---

* In this report, storage refers to the total volume of storage space available behind a dam at its completion. Some storage space may be occupied by sediment, and some by water; some space may be unoccupied at any given time in the history of the structure.

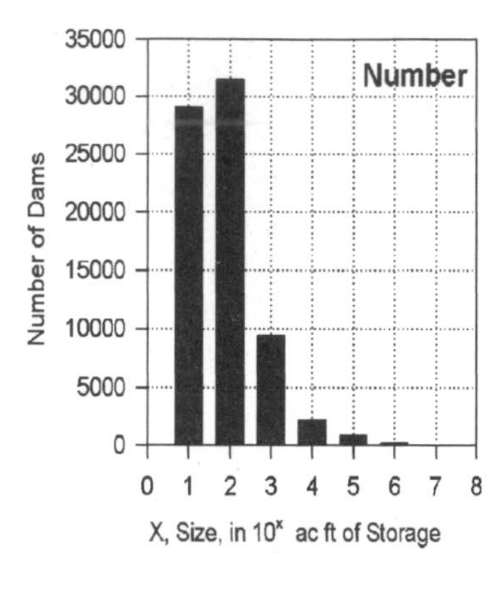

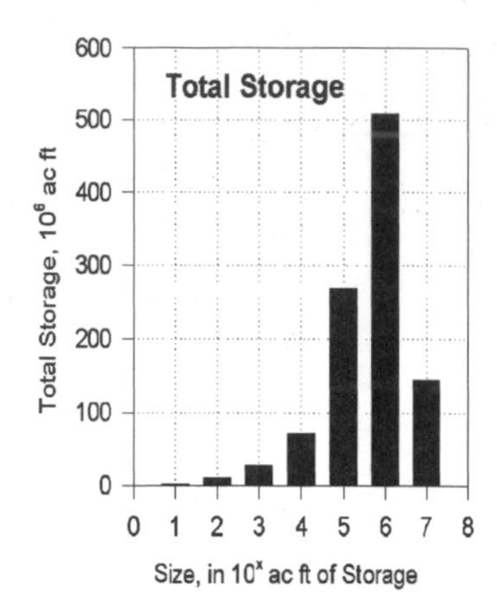

Figure 1.1 Most U.S. dams are small (chart on left), but the most storage volume is contained in reservoirs behind large dams (chart on right). *Sources*: Data from Federal Emergency Management Agency and U.S. Army Corps of Engineers (1996); calculations from Graf (1999).

structures create reservoirs or artificial lakes that impound water and release part of it through the dam on schedules determined by operators for various purposes (e.g., flood control, electric power generation). Reservoirs behind these dams experience large changes in water level annually. Dams that store no water or very little water are low structures across river channels designed either to raise the water level upstream for navigation purposes or divert flows into canal headings for distribution away from the stream (Figure 1.2). Some are run-of-river hydroelectric facilities. These low dams allow river discharges to flow over their crests, form reservoirs characterized by little fluctuation in water level, and do not generally affect moderate and high flows downstream. In dry land regions, diversion works often desiccate downstream areas by steering most or all of the low flows into canals. Some dams with storage reservoirs create a run-of-river downstream condition through operating releases, whereby the dam releases water at approximately the same rate as the reservoir receives it.

The difference between run-of-river structures and those that store significant and variable amounts of inflows from upstream is important from a physical and biological standpoint for reaches of the river downstream. If a dam is of the run-of-river type and does not divert a significant portion of the flow, then it does not alter the fundamental characteristics of

Figure 1.2 Forge Creek Dam in Cades Cove, in eastern Tennessee, is an example of a small, run-of-river structure. This view looks upstream from the west bank of the head-race and shows the main and diversion dams. A small diversion canal that supplied water to a mill is on the left. Courtesy of the Library of Congress, Prints and Photographs Division, Historic American Buildings Survey, Reproduction Number 5-CADCO, 1-24.

the flow of water downstream. Such a dam does not alter peak flows, mean flows, or low flows; does not change the timing or seasonality of peak or low flows; and does not alter the rate of change between high and low flows. Dams with storage reservoirs have the capability to effect such changes on downstream flow. Any problems in linking cause and effect become even more complex when attempting to predict the outcomes of dam removal. Because storage reservoirs have numerous and complicated effects when they are in place, their removal also is likely to produce complex changes in hydrology and downstream physical and biological systems.

From a design standpoint, the range of approaches to dam building seems endless. Local conditions, availability of building materials, and sophistication of the designers and builders are all highly variable, but there are a few standard types of structures that are most common: crib, earth fill, rock fill, concrete gravity, concrete arch, and concrete buttress dams (Jackson, 1988; U.S. Bureau of Reclamation, 1987).

Crib dams are especially common among older, small, run-of-the-river structures constructed as far back as colonial times in the United States. Cribbing constructed of timber forms an outer box for these low dams to create a linear barrier across the stream. The interior of the box often is filled with rocks for stability and sometimes further stabilized with wire or brush blankets (Figure 1.3). Because these dams often have constant

Figure 1.3 **Felix Dam, shown here in 1995, is a timber crib dam on the Schuylkill River in Pennsylvania. This dam was partially breached during Tropical Storm Floyd in September 1999.** Courtesy of the Pennsylvania Department of Environmental Protection.

overflow, they tend to deteriorate more rapidly than do some other types. As a result, many older crib dams have changed in form over the years, first built as wooden structures and later armored with a layer of concrete.

Earth fill dams, the most common general type of modern times, often are used as small storage structures. They are constructed from local earth materials that are shaped and rolled into a sill across the watercourse to be dammed. In cross section, along the alignment of the stream, the dam has a broad base with gradually sloping faces. All dams require spillway structures because, if the dam is overtopped by water flow, it is likely to be eroded and breached.

Rock fill dams use rock for weight and stability in association with a low-permeability membrane to provide watertightness. Like earth fill structures, rock fill dams are protected from destructive overflows by spillways, which drain off excess water when the reservoir approaches a full state (Figure 1.4).

Gravity dams consist of large masses of materials held in place by their own enormous weight (Figure 1.5). The construction material for modern gravity dams is usually concrete, but older structures often were

Figure 1.4 Township Line Dam, across Township Line Run in Pennsylvania, is an example of an earth fill dam. Courtesy of the Pennsylvania Department of Environmental Protection.

Figure 1.5. The Christian E. Siegrist Dam, constructed in 1993 across Mill Creek in Pennsylvania, is an example of a roller-compacted concrete gravity dam. Courtesy of the Pennsylvania Department of Environmental Protection.

built of masonry; cut and dressed stone blocks; or, in some eastern areas, brick. They usually are founded on a bedrock base and may be either linear or curved in plan. They are wider at the base than at the top to account for increased water pressure at the lower edges. Spillways or gates that permit the passage of water through, over, or around the structure to prevent overtopping often protect dams of this type.

Arch dams commonly are found where the dam site is a narrow constriction of the valley or canyon containing the stream. There are two subtypes: single and multiple. A single-arch dam spans the valley opening as one single structure and is anchored in the sidewalls by thrust blocks (Figure 1.6). In addition to a normal spillway, an emergency spillway is required to help prevent overtopping during high flows. A spillway may appear on the dam crest, gates in the structure may be used to drain excess water from the reservoir, or there may be bypass conduits that conduct water through the canyon or valley walls around the structure. The length of a single-arch dam usually is no more than about 10 times its height. Multiple-arch dams span valley openings that exceed this 10-to-1 ratio and have concrete arches connecting buttresses. A common design strategy in

Figure 1.6 Rindge Dam, built in the 1920s on Malibu Creek in California, is an example of a concrete arch dam. Courtesy of Sarah Baish.

the late 1800s and early 1900s was to make the connections between the buttresses with sloping, barrel-shaped arches of uniform thickness. Many water storage structures constructed at the turn of the twentieth century in western states are of this type. The majority of arch dams are concrete, although there are some masonry single-arch structures.

Buttress dams are made of flat decking that slopes from the crest to the base, usually with the decking inclined in the downstream direction (Figure 1.7). Numerous vertical buttresses anchored in bedrock support the decking, so the resulting structure is hollow rather than filled like a concrete or masonry gravity dam. Spillways are usually included to protect the basic structural integrity of the dams. Buttress dams commonly were constructed during the 1930s when labor costs were low relative to material costs; they were seldom built after World War II. The most common building material for buttress dams was concrete.

Ownership

The Federal Emergency Management Agency and USACE analyzed the NID to determine ownership of the 76,000 structures they recorded in

Figure 1.7 **Bear Creek Dam in Pennsylvania, shown here under construction in 1915, is an example of a buttress dam. This dam was breached in 1999.** Courtesy of the Pennsylvania Department of Environmental Protection.

the 1996 accounting. The analysis revealed that the majority of dams are privately owned (Table 1.1). Because there are many more small structures than other sizes, and because small structures are usually privately owned, private ownership is a major factor in the consideration of dam removal issues. Local government agencies own the next-largest share of the total inventory of U.S. dams, again largely concentrated in the small size range. Significantly, smaller proportions of the total inventory are the property of state agencies, the federal government, and public utilities.

The federal government owns only a small proportion of the total stock of dams in the nation, with its ownership concentrated among the largest structures. The significance of this observation is that the federal government owns the largest amount of storage capacity. Any removal decisions related to these very large structures would involve complex regional and national trade-offs among environmental, social, and economic concerns. Scientific issues and decision making are much more

Table 1.1 Ownership of American Dams

Owner	Number	Percentage of Total
Private	43,661	58.1
Local	12,859	17.1
State	3,680	4.9
Federal	2,209	2.9
Public utility	1,659	2.2
Undetermined[a]	11,119	14.8
Total	75,187	100.0

Source: Data from Federal Emergency Management Agency and U.S. Army Corps of Engineers (1996).
[a] Abandoned or of questionable ownership.

complex and difficult to resolve for these large federal structures than for the smaller privately and locally owned ones.

DISTRIBUTION

Dams are a component of the American landscape. They appear in every major and minor river system of the lower 48 states and are found in every county and territory of the nation. Texas has the most dams of any state, almost 7,000, and Worcester County, Massachusetts, has the most of any county, 425 (Federal Emergency Management Agency and U.S. Army Corps of Engineers, 1996). The greatest concentration of dams is in the southern, midwestern, and Plains states (Graf, 1999). Fewer large dams are located in the interior western regions because of lower population density and lack of water (Figure 1.8).

A map of the water volume stored behind the dams would look different from the map of dam density because many of the large water-storage structures are in the West and Great Plains areas. Thus, the downstream effects from a disruption of the hydrologic system are likely to be greatest in those areas. Smaller, run-of-river structures with little or no storage are common in Atlantic coastal areas and the Midwest; the major environmental issues connected with dams in those regions are likely to be related to the disruption of fish passage rather than flow regulation.

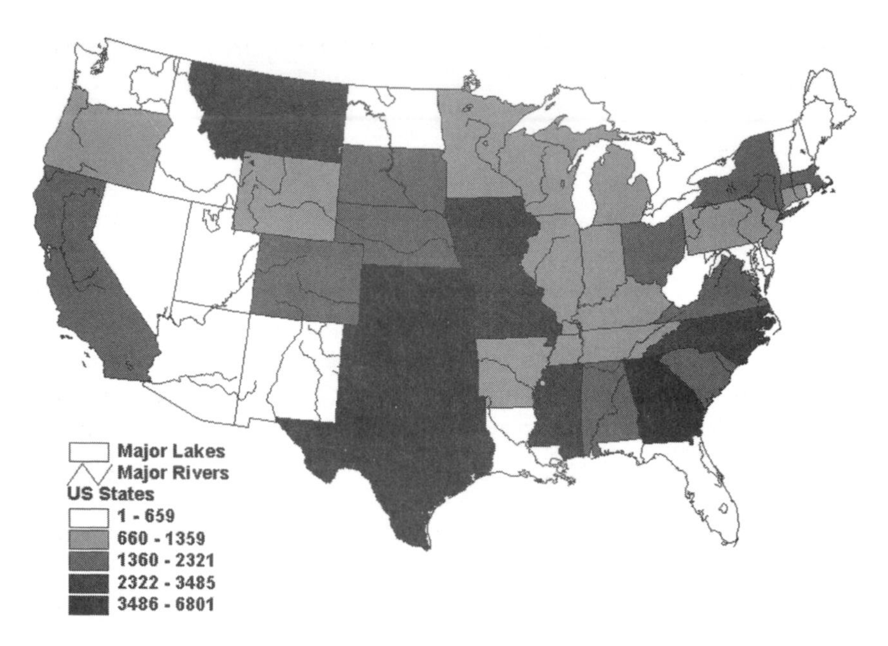

Figure 1.8 This map shows the distribution of existing American dams, with the higher densities indicated by the darker colors. *Sources*: Data from Federal Emergency Management Agency and U.S. Army Corps of Engineers (1996); map from Graf (2001a).

REASONS FOR DAM BUILDING

Dams have been part of the American infrastructure from prehistoric times.* In the eastern and midwestern regions and the Pacific Northwest, Native Americans constructed low dams and fish weirs. In drier western and southwestern areas, extensive irrigation works supported agriculture for the continent's first cities, pueblos with human populations numbering many thousands. It was European settlement and technology, however, that initiated the construction of permanent dams that exerted control over river hydrology. Dams diverted stream flow to power mills throughout the 13 original colonies and in southern coastal areas to water rice and indigo crops. The oldest surviving dam is Mill Pond Dam in

---

*The ideas expressed here were derived from research supported by a National Science Foundation Grant to W. L. Graf.

Newington, Connecticut, built in 1677; nationally, more than 20 dams survive from the 1700s (Federal Emergency Management Agency and U.S. Army Corps of Engineers, 1996).

The twentieth century saw the construction of more than 80 percent of all the existing dams in the nation. As population growth, expanded agriculture, and industrialization increased the demand for water control infrastructure, the nation invested in building dams of all sizes. During the twentieth century, the amount of total storage behind dams grew from a relatively small amount to almost 1 billion acre-feet (Figure 1.9). Although some very large structures were products of New Deal or World War II construction, the great dam building era in the United States was from about 1950 to about 1970. The peak construction year was 1960, with more than 3,000 dams completed in a single year. The decade of the 1960s saw the construction of more than one-quarter of all the structures existing as of 1996 (Federal Emergency Management Agency and U.S. Army Corps of Engineers, 1996). After about 1980, the installation of new dams dramatically declined, partly because of increased public scrutiny and environmental concern, but also because almost all of the geotechnically desirable dam sites had been used.

This historical account of dam building is of more than academic interest. Because the majority of dams were built in the mid-twentieth century, they are of great technical, policy, and scientific importance now, 50 years later, for three reasons. First, many of the small and medium-

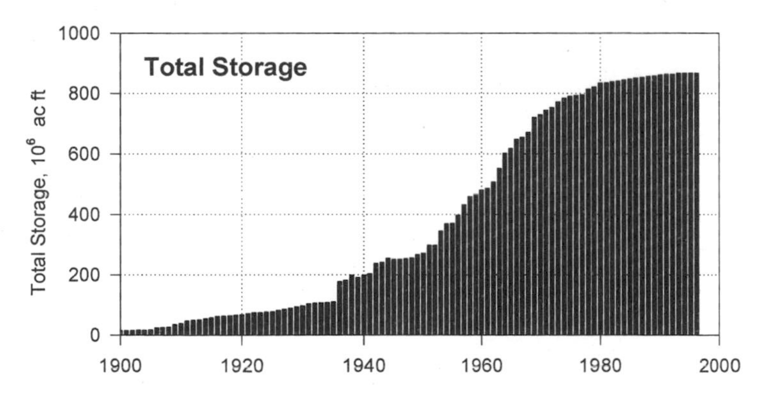

Figure 1.9 Reservoir storage behind American dams increased steadily during the twentieth century. *Sources*: Data from Federal Emergency Management Agency and U.S. Army Corps of Engineers (1996); calculations from Graf (1999).

sized dams constructed during this period (the vast majority of all dams constructed then) have design lives of about 50 years, so many are now in need of repair. Second, some small, privately owned hydroelectric dams constructed on public waterways were constructed under licenses from FERC and its predecessors, and those licenses were for 50-year periods. At the end of the license term, a comprehensive reevaluation is to be performed of the environmental and developmental aspects of the project for an extended operating license. Third, in many cases it has taken decades for the environmental effects of the dams to become obvious. Significant time has been required for their influence on aquatic and riparian systems to play itself out and become manifested in changes in biological systems.

In current discussions about the removal of dams, considerable attention is given to the detrimental effects of the structures to the environment and/or costs that were either unknown or ignored when the dams were built. However, every dam was based on a perception that its benefits were real and tangible, and the resulting infrastructure provided by dams indeed has generated benefits for the nation. The question of whether individual structures need to be removed can be assessed most realistically based on an understanding of the reasons for the original dam building. The most common reasons for dam construction include recreation, water supply for fire control and farm ponds, flood control, water supply, irrigation, waste disposal, electricity production, and navigation (Table 1.2).

Recreation

Many reservoirs were created for recreational reasons. Flat-water recreation is a significant component of regional economic activity, especially in the southeastern, midwestern, and Plains states. Nationwide, more than 27 million people are power-boaters, and their more than 16 million craft dot the nation's reservoirs (National Sporting Goods Association, 1998). Recreational boaters use reservoirs of all sizes as well as the elevated levels of rivers controlled for commercial rafting use. Fishing in reservoirs is very popular, a $28 billion-a-year industry in the United States (American Sportfishing Association, 2001) (Figure 1.10). Releases of cold water from medium-sized and large reservoirs also support many trophy trout fisheries that otherwise would not exist.

Table 1.2 Primary Purposes of American Dams[a]

Primary Purpose	Number of Dams
Recreation	26,817
Fire and farm ponds	12,532
Flood control	10,971
Water supply	7,293
Irrigation	7,223
Tailings and waste	6,756
Hydroelectric	2,259
Navigation	226
Undetermined	1,110
Total	75,187

Source: Data from Federal Emergency Management Agency and U.S. Army Corps of Engineers (1996).
[a] Many structures are multipurpose.

Fire and Farm Ponds

Fire and farm ponds are common in rural areas across the United States. Since colonial times, farmers have been building dams and creating small reservoirs to impound water for livestock or agricultural uses (Forest Preserve District of Cook County, 1971). These ponds also traditionally have supplied water in case of fire. For the past two decades, a federal program run by the U.S. Soil Conservation Service (now the Natural Resources Conservation Service) has stimulated construction of these types of ponds (Forest Preserve District of Cook County, 1971). Today, most of these fire and farm ponds have multiple uses, including recreation such as swimming, fishing, and ice-skating. In Iowa alone, farm pond owners host 1.6 million fishing trips by licensed anglers each year (State of Iowa, Department of Natural Resources, 2002).

Flood Control

Flood control is a major function of large, multipurpose dams in all parts of the nation, but especially in the East and Midwest. Medium-sized and large dams are used for flood control because large volumes of storage are

Figure 1.10 **A young angler displays his catch taken at a lake located behind the Center Hill Dam on the Cumberland River in Tennessee. Recreational fishing is a benefit derived from most reservoirs.** Courtesy of the U.S. Army Corps of Engineers.

required to capture potentially hazardous runoff and store it for subsequent gradual release. The large volumes of water involved are available only episodically, but this is sufficient to provide for needs such as drinking water supply and electric power production. The USACE is authorized by the U.S. Congress to construct dams largely for flood control purposes. The nation has invested more than $3 billion in flood control works, yet annual losses from floods continue to increase (National Weather Service, 1999), and floods account for more loss of life than any other natural hazard in the United States (U.S. Census Bureau, 1999).

Water Supply

The need for water for urban, domestic, and industrial use spurred the construction of many dams, ranging from small, run-of-river structures diverting stream flow into distribution systems to medium-sized and large structures providing temporary storage. Withdrawals of water from the nation's streams and rivers amount to about 183 gallons per person per day, and industrial uses demand 24 billion gallons per day. About 60 percent of the domestic water and 80 percent of the industrial supply comes from surface water in streams and rivers (Solley et al., 1998). Water is removed from the riverine part of the hydrologic cycle by diversion or storage in reservoirs. Some of this water is returned by wastewater and treatment plant discharges. These are significant diversions from an ecological perspective, disrupting the riverine part of the cycle and important components of the overall social and economic system. Economic development in the United States has caused water consumption to double in the past 40 years, and increases of this scale are likely to continue (Postel, 2000).

Irrigation

Irrigation diversions for agriculture are common functions of low dams in the plains and western states, where rainfall is not consistent enough for the production of crops. Medium-sized and large dams create storage reservoirs in the upper portions of watersheds, filling them with runoff and snowmelt in winter and spring and releasing water for downstream diversions into lateral distribution systems during the growing season. Irrigation systems withdraw 134 billion gallons per day from the nation's streams and rivers (Solley et al., 1999), and, unlike the consumptive process that occurs in many industrial uses, most is consumed by evapotranspiration. The U.S. Bureau of Reclamation was authorized by Congress to build large structures in western states for irrigation water supply; many smaller structures are the products of private investment.

Waste Disposal

The construction and maintenance of dams is required for waste disposal associated with several activities, particularly mining and industrial animal husbandry. The processing of mineral ores produces large quantities of liquid waste, slurry, and tailings, which contain water, acids, and sediment. The containment of these materials, usually by small dams, allows the water to evaporate, producing solids, which are more easily managed. Similarly, extensive industrialized production of chicken, pork, and beef results in the generation of large amounts of animal wastes, which often are retained in ponds by small dams before further treatment. Waste disposal ponds and their dams pose special problems because of the need for hazardous biomaterials management in their operation and removal.

Waterpower

Waterpower was the primary reason for the construction of many of the older dams in the United States. From colonial times until the late 1800s, water diverted from streams hydraulically drove machinery to grind grain and produce goods ranging from textiles to sawn lumber. Most of these early structures were small, but they dotted coastal rivers, particularly in the eastern states bordering the Atlantic Ocean and in midwestern states. The advent of steam power made many of these structures obsolete for their original intended purpose, but some were refitted for other purposes, including the production of electricity. Only 3 percent of all dams (2,259) are hydroelectric (Table 1.2).

Electricity Production

Hydroelectric dams use waterpower to turn turbines (Figure 1.11). Water may simply be passed through the structure of a dam to generate electricity, or it may be diverted through canals and pipes to off stream locations for this purpose. Electricity production therefore involves dams of all sizes, ranging from low-diversion works to large storage structures. Hydroelectric plants may produce base-load power by releasing consistent amounts of water to the turbines, or they may produce peak power by releasing water on a schedule to coincide with maximum demand. Dams produce about 10 percent of the nation's electricity and about 70 percent of electrical power in the Pacific Northwest (U.S. Census Bureau, 1999).

Recent problems with electricity production and distribution, notably in California during the winter of 2000–2001 and to a lesser degree in the summer of 2001, may influence decisions about hydroelectric dams. A heightened appreciation for electricity from such structures (inexpensive, reliable, low emissions) means that additional care needs to be taken when considering their removal. However, there are three reasons why increased concern about hydropower is not likely to change the current decision-making processes with respect to removal of some structures. First, as outlined in the next major section of this chapter, the majority of decisions to remove dams at present revolve around concerns over safety or obsolescence, and power production is only a minor consideration. Second, almost all of the dams being removed or likely to be con-

Figure 1.11 Bonneville Dam, on the Columbia River in Oregon, is an example of a large hydroelectric structure. It was completed in 1937 and produces 6 million megawatt-hours of electricity, enough for more than 500,000 homes. The dam may also contribute to declines in the salmon fishery. Courtesy of the U.S. Army Corps of Engineers.

sidered for removal in the near future are small, run-of-river structures, which generally produce little or no electricity. Third, medium-sized hydropower dams considered for removal are likely to be removed regardless of the electricity market, because of other factors driving the removal process. Condit Dam on the White Salmon River of Washington State, for example, has a reservoir that is filled largely with sediment, which reduces the effectiveness of the structure for power production.

Navigation

Navigation on the nation's inland rivers depends on lock and dam systems that maintain pools of water deep enough to accommodate boat and barge traffic. About 25,000 miles of the nation's streams support transportation of goods, with their water levels regulated by dams. Small dams raise water elevations, with boat and barge passage to and from various sections provided by locks adjacent to the dams. In upper reaches of watersheds, large storage dams impound reservoirs that release water to sustain the downstream pools. Direct operating revenues of the system are more than $3 billion per year, accounting for the annual transport of 622 million tons of cargo (U.S. Department of Transportation, 2001). Although numerous industries depend on this transportation system, bulk cargos of grain, coal, and oil are most common.

REASONS FOR DAM REMOVALS

Just as Americans have been builders of dams, they have also torn some of them down. Some owners expected that the structures would not be permanent, and when the original purpose was served, they removed the dams. Dams built to serve sawmills in remote forests, for example, diverted stream flow to power the saws, and after the lumber was harvested, the mills and their dams disappeared from the landscape. The demise of many eighteenth-century gristmills in New England resulted in the removal of their diversion works; similarly, the waterworks associated with nineteenth-century mountain mining areas in the West also have disappeared. The removal of dams for purposes of restoring and maintaining some measure of environmental quality is a more recent phenomenon, but there is much national experience with the basic concept of dam removal.

Dam removal is one option for dealing with the effects of dams that are detrimental to environmental quality. However, progress has been made in the past two decades towards mitigating the undesirable physical and biological effects of dams while preserving the functional objectives (when still viable) of a dam and its impoundment. Most of these advances have resulted from actions related to the Federal Energy Regulatory Commission's licensing of privately owned hydropower projects and environmental assessments for federal projects done in response to NEPA requirements. FERC licensing procedures and NEPA environmental impact assessments embody some of the collaborative processes recommended in this report for evaluating dam removal (EPRI, 2000a; NHA, 1999; EPRI, 1996).

Advances to protect physical and biological processes have been made in the following areas:

- Fish passage (upstream and downstream) (Odeh, 1999 and 2000; EPRI, 1998; Clay, 1995; USDOE, 1994 and 1991; AFS, 1993; Mattice, 1991; AFS, 1985)

- Fish entrainment protection (EPRI, 2001; Coutant, 2001; Odeh, 2000 and 1999; EPRI, 1998; FERC, 1995; AFS, 1993)
- Instream flow protection (EPRI, 2001b; USDOE, 1991; AFS, 1985)
- Water quality protection (EPRI, 2002; USDOE, 1991; Mattice, 1991)
- Sediment management (EPRI, 2000c; White, 2001)
- Riparian area protection and management (EPA, 2001)

More information on mitigation measures at specific hydropower projects can be obtained from the FERC Library in Washington, D.C., or from the FERC Records and Information Management System (RIMS). RIMS can also be accessed via the World Wide Web at http://rimsweb1.ferc.fed.uc/rims.q?rp2~intro. This body of information is particularly important to owners, operators, and other stakeholders involved in dam removal decisions when alternatives to removal are preferred. In some cases, however, owners determine that dam removal is an appropriate option, and just as it is true that there were definable reasons for installing the dam, so there are definable reasons for removing it.

Structural Obsolescence

A major expense associated with maintaining aging dams is the cost of structural repair required in the course of normal dam operations. Many dams have a useful life expectancy of about 50 years (River Alliance of Wisconsin and Trout Unlimited, 2000). This life expectancy typically is used in economic analyses related to dams. Maintenance and upgrading may extend this life span, and poor maintenance or abandonment may reduce it. Theoretically, dams could last forever if properly maintained. Of the entire formal list of dams maintained by the USACE, more than 22,000 (30 percent) are already more than 50 years old, and by 2020 more than 60,000 (80 percent) will be more than 50 years old (Federal Emergency Management Agency and U.S. Army Corps of Engineers, 1996). Because of concerns that "thousands of U.S. dams built in the 1930s and 1940s are nearing the end of their design life," the American Society of Civil Engineers developed a set of guidelines and principles for the retirement of dams, including their removal (Task Committee on Guidelines for Retirement of Dams and Hydroelectric Facilities, 1997). The resulting guidance document contains descriptions of techniques, methods, and procedures for dam removal and includes numerous case studies.

Many dams require substantial overhaul after several decades of continuous operation. Run-of-river structures have water continually pouring over their crests, and erosion of exposed parts is inevitable. All dams leak to some degree, and the water passing through them often leaches calcium carbonate from the cement and mortar, resulting in reduced cohesion and disintegration. Erosion of the riverbed below some dams results in a gradual undermining of the structure; in some cases, weakened abutments and anchors require refurbishing. Without proper maintenance, structural deterioration can lead to a dam's collapse (Figure 1.12). Dams that were installed 50 or 100 years ago may require substantial investments to return them to safe, modern operating condition. In many cases, if the owner is an individual or small business, removal is the only reasonable, economical alternative. In other cases, the dam is abandoned or orphaned.

Safety and Security Considerations

Dam safety and security is a major issue in the consideration of dam removal. The legislation that established the National Dam Safety Pro-

Figure 1.12 **The 1933 collapse of Castlewood Dam on Cherry Creek in Colorado is an example of the result of unsafe dam construction and maintenance.** Courtesy of the Denver Public Library, Western History Collection; photograph by Harry Rhoads.

gram and increased public concern about dam safety dictates due care by every dam owner in the country to ensure each dam's safe operation. Dam failures inundate downstream areas with unexpected floods and disastrous results. Historical dam failures in the United States include South Fork Dam upstream from Johnstown, Pennsylvania. The dam collapsed during an 1889 storm, and the ensuing flood killed 2,209 people (McCullough, 1968). The infamous Saint Francis Dam on the Santa Clara River of Southern California killed 525 people because of its collapse in 1928 (Garrison, 1973). In the 1970s, four dam collapses (Buffalo Creek, West Virginia; Canyon Lake, South Dakota; Teton, Idaho; and Kelly Barnes, Georgia) took 300 lives and initiated modern dam safety efforts.

Since 1980, the loss of life from dam failures has declined. However, environmental changes resulting from dam breaches continue to cause problems. In the southeastern states, hurricanes in the 1990s triggered river flooding that breached small waste retention dams and spread animal waste, primarily from hog farms, throughout downstream areas (Schwab, 2000). Breaching of mining tailings ponds during floods is also a problem that plagues some western states.

The downstream hazard posed by a dam depends on the physical condition of the dam, downstream river channel geomorphology, and distribution of the human population downstream. Dams in structural disrepair and populations living or working in flood-prone locations increase the hazard. Engineers have concluded that more than 13,000 dams in the NID pose significant hazards (a risk of property damage if the dam fails), and 10,700 are high hazard risks (with the potential for loss of life if the dam fails). Taken together, these dams constitute about 32 percent of all the dams in the inventory. In some cases, the owner of a dam in one of these risk categories may find it easier to simply remove the structure than to mitigate the risks from its continuing operation (Figure 1.13).

Economic Obsolescence

Most dams that have been removed from U.S. rivers have outlived their economic usefulness. In many cases, the reasons for their initial construction no longer apply. Many of the dams that diverted eastern streams for millraces or raised river levels to drive waterwheels lasted longer than the mills they served. In other cases, early hydroelectric facilities became antiquated with the development of regional power grids fed by larger, more

Figure 1.13 **The 56-foot-high Bluebird Dam, located in Rocky Mountain National Park, Colorado, was removed in the summers of 1989 and 1990 after the collapse of a nearby dam killed three people and raised safety issues.** Courtesy of Rocky Mountain National Park.

efficient sources of electrical energy. Even if the dam in question no longer produces income, expenses continue to accrue, including maintenance and insurance costs. The removal costs for small, run-of-river structures in the upper Midwest typically run about $100,000 or less.* This may be much less expensive than retooling a dam for a new purpose or performing needed structural repairs.

Dam owners may choose to remove a dam to eliminate their own potential liability. Small, run-of-river structures in humid regions of the country have river flow continually pouring over their crests, creating a hydraulic jump. The resulting turbulence and reverse eddies that sometimes result can be deadly traps for boaters and canoeists. People fishing from dams and related structures risk serious injury or drowning. The liability of the dam owner in such a case of injury or death is unclear, but some owners prefer to avoid the risk by removing the structure. In the upper Midwest, owners of small dams report insurance premiums of several thousand dollars per year, an expense that is eliminated

* This figure is based on reports by state officials at a December 2000 short course on dam removal at the University of Wisconsin.

by removal of the structure. The threat of liability for injuries or property damage following a dam collapse gives dam owners an economic incentive to repair or remove unsafe dams, and removal may be cheaper than repair.

In the wake of the terrorist attacks of September 11, 2001, FEMA expressed an increasing awareness of the vulnerability of dams to security threats. This event may provide new incentives for either more attention to the security of dams or renewed interest in their removal.

Recreational Opportunities

Dams and their reservoirs make flat-water recreation possible; dam removal, although it eliminates reservoirs, often changes and sometimes improves recreational opportunities downstream. White-water boating in canyon rivers is enhanced by more natural river flows. Some dams operated as hydropower facilities create rapidly changing conditions for rafters as dam operators produce peak flows (Box 1.1). In flatland streams, canoeists and boaters seek continuous uninterrupted lengths of river, and campers and others enjoy stable streamside areas with natural forests. Sport fishing, especially for trout in eastern and midwestern streams, benefits from rivers without subdivision by dams. Such fishing also is enhanced by the wide variety of habitats that results from unregulated rivers and their flows. However, dams have supported trout fisheries that would not exist without the coldwater releases from reservoirs. In addition, reservoirs have provided habitat for largemouth bass, a fish prized by anglers.

Water Quality and Quantity Issues

Dam removal affects water quality and quantity because many reservoirs created by dams provide drinking water for human consumption. Given the importance of clean drinking water, the impact of dam removal needs to be considered carefully. A reservoir used for drinking water may be the pristine source of water for the region. If the dam is removed from such a reservoir, the nearby population may have to turn to groundwater supplies, which may be more contaminated and more expensive to obtain. On the other hand, the removal of a dam may improve water quality in at least two

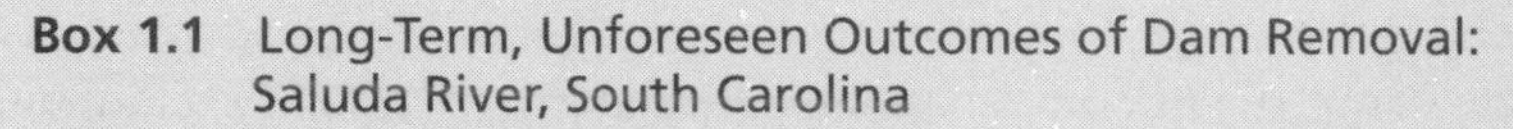

Box 1.1 Long-Term, Unforeseen Outcomes of Dam Removal: Saluda River, South Carolina

Mill Race Rapids in 2001

Courtesy of William Graf

The need for extensive consideration of a variety of perspectives when considering a dam removal is illustrated by the experience on the lower Saluda River in South Carolina. A milldam was removed several decades ago using explosives. The remaining jagged bedrock, fractured by the removal process, contributes to the formation of rapids at the former dam site (Holleman, 2001). The rapids are situated on a ledge of metamorphic rock, and pieces of the former dam become exposed at the surface of the stream at low water levels. Uses of the river reach known as Mill Race Rapids include canoeing and kayaking, and the former dam site is a favored location for fishing, wading, and swimming.

Mill Race Rapids is about 9 miles downstream from Murray Lake Dam, a medium-sized structure that produces hydroelectricity. The operation of Murray Lake Dam (officially known as the Dreher Shoals Dam) causes rapid fluctuations in discharge, and people often are stranded on the rocks by rapidly rising water levels. In 2001, a swimmer trying to rescue a stranded wader drowned after he was wedged into the rocks and submerged by rapidly rising waters. Kayakers describe the rapids as simply "dangerous all the time" (Holleman, 2001). The owners of Murray Lake Dam have installed warning lights and sirens to alert recreational users to impending rises in discharge, and the South Carolina Department of Natural Resources patrols the river. Despite these efforts, four people have died in the rapids of the lower river in the last seven years.

The experience of Mill Race Rapids provides a general lesson for decision makers in other dam removal cases. When the milldam was removed, the prospect of such problems was not publicly discussed. Given the pressure on river resources from recreational users, present-day plans for any dam removal need to account for use of the site by a variety of people long after the dam has disappeared.

important ways: by increasing the amount of dissolved oxygen and by returning water temperature to natural conditions. Water quality considerations in dam removal decisions are important from a regulatory standpoint because the stream that results from a dam removal is subject to evaluation against standards imposed by the Clean Water Act. Dam removal also may release accumulated sediment. This sediment, whether toxic or not, can reduce the quality of downstream water for human consumption, but may also restore stream habitat through deposition. However, this effect is usually temporary but can raise water supply treatment costs.

Dam removal also may affect groundwater supplies. For example, in central Vermont, FERC decided not to order dam removal in a voluntary surrender of a license. Instead, the commission allowed a local village to buy the dam. The village depended on this reservoir to maintain a high water table and previously had experienced a water shortage when another dam on the same river was removed (Pyle, 1995).

Ecosystem Restoration

When a dam is removed, the river course once inundated by reservoir waters is restored. In addition, river reaches downstream from the removed dam also may be restored to a more natural condition. In their natural conditions, rivers are highly integrated ecological systems. Dams fragment the networks into isolated bits and pieces that are biologically and physically separated from each other. The principal removal efforts to date involve dams that fragment streams and block salmon spawning runs. Several Pacific Northwest dams are candidates for removal for this reason. This region is famous for severely depleted salmon runs and large hydroelectric projects that may be contributing to the declines. Even resident or native river fish often have wide annual ranges when not blocked by dams. Aquatic organisms often are prevented from reaching their original natural range in regulated and dammed rivers, so dam removal is an obvious method of reconnecting the system. The state of Wisconsin, for example, is removing four dams on the Baraboo River with the intention of restoring the connected system that once existed there (Figure 1.14), and many removals of small dams on Atlantic Seaboard streams seek to reestablish access to spawning areas for anadromous fishes.

The objectives of a dam removal need to be articulated before initiation of the project. To many, the recovery of a river implies that the

Figure 1.14 Waterworks Dam on the Baraboo River in Wisconsin was removed during the winter of 1997–1998. Courtesy of the River Alliance of Wisconsin.

physical and biological components will return to the same level that existed before the building of the dam. Rarely is this possible, because of the other impacts and changes that have taken place in the watershed. Rehabilitation implies that the physical and biological processes that define the river are returned to a functional level. This level is determined by the input from the upstream river, localized inputs, and location of the rehabilitated reach in reference to the rest of the watershed. The actions taken to achieve rehabilitation, and perhaps eventual recovery, of a rivers' ecosystem are known as restoration activities.

Similarly, many assume that the post-removal recovery of rivers and the species they support will be self-sustaining and not require additional actions by people. This may not be the case. Watersheds are dynamic and continuously respond to impacts and changes. The placement of a dam in a river fragments not only the river, but also the watershed. The removal of a dam will not automatically result in the full recovery of the river or the species that it once supported. It is essential to evaluate each dam removal in the context of other community issues and the location of the dam within the watershed.

Ecosystem restoration may be the most controversial rationale for dam removal. Many people view reservoirs as normal and natural components of the ecosystem and worry that any change back to the original natural river ecosystem will destroy existing wildlife and fish habitats. Many people prefer reservoirs to rivers, enjoy power boating more than white-water rafting, and would rather fish for bass and catfish than salmon and trout. Such a bias in favor of existing, anthropogenic environments, combined with common personal recreational preferences, are conflicts that are often difficult to resolve. The importance of restoration to the public is subject to changing value systems. Fifty years ago, when many river dams were under construction, the restoration of aquatic ecosystems was virtually unheard of; now it is a national policy articulated by the preamble of the Clean Water Act.

DAMS REMOVED IN THE UNITED STATES

Number of Dams Removed

Data on dams that have been removed are difficult to obtain. Many removal projects left little evidence of the former structure on the landscape, and even less documentation. Scientific evaluations of these removals are almost nonexistent. Research has been limited because the scientific community only recently has recognized such work as interesting from a scientific as well as policy standpoint. Additionally, funding for such research has not been readily available. The most extensive recent effort to collate information about dam removals was carried out by nongovernmental organizations (American Rivers et al., 1999). Their accounting identified 467 structures that had been removed; the total continues to rise as more examples are identified. In another project, the National Performance of Dams Program at Stanford University is compiling a database on removed dams. In December 2000, officials of the Wisconsin Department of Natural Resources informed participants in a university short course that 50 dams had been confirmed as removed in that state. Further research has turned up data on 120 more, and Wisconsin officials suspect that as many as 500 dams have been removed in the state. Similar undercounting is likely throughout the nation.

The physical removal of dams can be undertaken using a variety of methods. Mechanical dismantling of the structure and physical removal of the debris usually begins with a breach to drain water stored behind the dam. For small, run-of-river structures, demolition of the remaining structure then can proceed while dealing with relatively shallow water conditions. For larger dams with significant storage, a systematic process of creating increasingly large notches in the structure results in a gradual drawdown of stored water. Although different rates of notching, drawdown, and removal are likely to be reflected by different responses in downstream sediment dynamics, little is known about these issues. Experimental drawdowns can provide information about the redistribution of sediments and channel changes in the newly exposed reservoir areas upstream from the dam and associated changes downstream.

SIZES OF DAMS REMOVED

Almost all of the dams removed thus far have been small ones, with storage capacities less than 100 acre-feet, along with a few medium-sized structures with storage capacity measured in thousands of acre-feet. Many of the structures removed were less than 30 feet high, but there have been exceptions. Sweasey Dam, removed from the Mad River in Northern California, was 55 feet high; Mississippi River Lock and Dam Number 26 was 98 feet high; and several industrial dams removed from Tennessee streams were more than 100 feet high. Size is not necessarily an indicator of the impact of removal, however, because the removal of small dams, especially low-head, run-of-river dams, can have substantial environmental benefits in opening fish passage and restoring ecosystem function to extensive river networks. Although Quaker Neck Dam on the Neuse River of North Carolina was only 7 feet high, its removal in 1998 opened more than 1,000 miles of river habitats to anadromous fish.

TYPES OF DAMS REMOVED

Although there has been no strict accounting of the types of dams removed, a review of available evidence and discussions with state dam officials in many areas show that the majority of structures removed from American streams have been low head, run-of-river dams with crib or rock fill structures, sometimes with coverings of concrete. A few larger, concrete arch structures are being considered for removal, such as Matilija Dam on Matilija Creek in Southern California (Box 1.2). This structure is 163 feet high and 620 feet long and impounds 1,800 acre-feet of water.

Box 1.2 Removing Matilija Dam

Fifty-year-old Matilija Dam, located on a tributary of the Ventura River in Southern California, in the next few years may become the largest dam ever removed in the United States. Many local, state, federal, and private organizations are working together to complete various studies and find the financial support needed to remove the obsolete structure.

Several federal agencies have conducted or are studying various aspects of the dam's possible removal. The U.S. Bureau of Reclamation and U.S. Army Corps of Engineers, for instance, are studying the cost and feasibility of removing the dam. The U.S. Geological Survey is studying the impact the removal might have on sensitive species. Since June 1999, Ventura County officials have agreed to support Matilija's removal, and in 2000, they removed a chunk of concrete that was 8 feet high and 30 feet long during a dam demolition demonstration project (Matilija Coalition, 2000). Nonprofit environmental organizations also actively support the removal. Proponents of removing the Matilija Dam include: The American Fisheries Society, American Rivers, California Trout, Environmental Coalition of Ventura County, Environmental Defense Center, Friends of the River, Friends of the Ventura River, Patagonia Inc., Trout Unlimited, and the Ventura County Chapter of the Surfrider Foundation. These groups advocate removing the dam to restore the habitat of the endangered southern steelhead and to allow the river's flow to replenish the sand and sediment of southern California's beaches (Environmental Defense Center, 2000). In 2000, the Matilija Coalition was founded to increase public awareness and secure the funding and congressional support necessary to ensure that Matilija Dam will be removed. Prominent individuals, such as Ed Henke, a former San Francisco sports figure who grew up along the Ventura River in the 1930s and 1940s, are also particularly outspoken proponents of the removal.

Ownership of Dams Removed

Almost all the dams removed in recent decades have been privately owned. In some cases, such as Edwards Dam on the Kennebec River in Maine, the original owner reached an agreement with other legal entities, and the public assumed ownership for the express purpose of dam removal. This series of events also has taken place on occasion in Wisconsin, where it is easier administratively for the state to assume ownership of the structures to be removed. In a considerable number of cases, dams that are targeted for removal by state game and fish departments as habitat restoration efforts do not have any known owner. These orphan dams truly become wards of the state, which then has to orchestrate their removal. In a similar example, Elwha and Glines Canyon dams on the Elwha River in the state of Washington started out as privately owned hydropower dams but now are owned by the federal government, and Congress has authorized their removal.

DISTRIBUTION OF REMOVALS

A geographical assessment of documented dam removals (Figure 1.15) as published by American Rivers et al. (1999) reveals a national pattern that is very different from the pattern of dam building (see Figure 1.8, p. 32). The states with the most dam removals are Pennsylvania, Ohio,

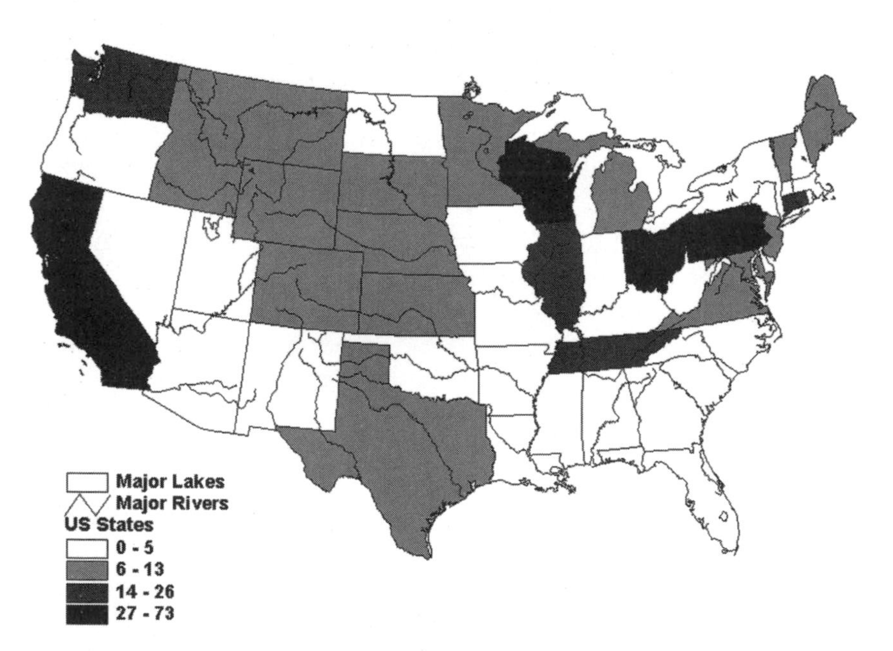

Figure 1.15 The map shows the distribution of dams removed in the United States, with the darker colors indicating the states where most removals occurred. *Sources*: Data courtesy of American Rivers et al. (1999); map designed by Graf (2001).

Wisconsin, and California. These are all states with some governmental commitment to providing administrative support for the activity.

Pennsylvania has an interest in reconnecting the Susquehanna River system, which drains into Chesapeake Bay. Because the state is part of a regional compact to enhance the bay's environmental quality, dam removal fits within a more general state policy goal. The criticality of connected river segments for the health of the bay provides an environmental incentive.

Wisconsin has a long history of fostering sport fishing, and in many cases the removal of antiquated crib and earth or rock fill dams that are small, run-of-river structures advances the state's general interest in improving aquatic habitat and supporting recreational fishing. Wisconsin also has paid considerable attention to reconstructing channels in previously inundated reservoir areas. California has environmental policies that stimulate the dam removal process. Some reservoirs are filled with sediment, so that the original purposes of the dams no longer can be served. In any case, the national distribution of dam removals is largely a function of local forces at work, combined with individual dam owners seeking to remove structures that they no longer want.

STATUS OF SCIENTIFIC RESEARCH ON DAM REMOVALS

Scientific research into the effects of dam removal is in its initial stages, and elaborate theories on the subject are not yet developed. Although investigations into the effects of dam installation have been ongoing for more than two decades, they are few in number and, as of late 2001, incomplete. In many cases, investigations of dam removal have not been reported in the refereed scientific literature. Examples include an evaluation of the possible removal of Searsville Dam in San Mateo County, California, being carried out by David Freyberg of Stanford University (Box 1.3). Matt Kondolf of the University of California at Berkeley is assessing the possible effects of removing Matilija Dam near Ventura, California. Employees of several federal agencies, including Brian Winter of the National Park Service and Tim Randle of the Bureau of Reclamation, are considering the outcomes of removing Elwha and Glines Canyon dams on the Elwha River in Washington State. Pat Shafroth and his associates in the Biological Resources Division of the U.S. Geological Survey (USGS) are involved in extensive investigations into the effects of dam

Box 1.3 Ongoing Management Studies and Research Projects: Searsville Dam and Lake Jasper Ridge Biological Preserve in California

Investigations at Searsville Dam near Menlo Park, California, illustrate the range of research initiated at the prospect of a dam removal decision. Searsville Dam was constructed in 1892 on San Francisquito Creek to provide water for various purposes during the dry months (Softky, 2000). The dam, constructed of interlocking concrete blocks, is 68 feet high. The dam is a source of concern to Stanford University, which has owned the dam since 1919, and the nearby community because of its negative environmental effects on the watershed. Problems include a buildup of sediment in the reservoir, upstream flooding, presence of exotic species, and impaired migration for endangered steelhead trout.

Searsville Dam and the associated lake are the focus of a number of ongoing management studies and research projects. The management issues facing the university as it makes decisions about the operation of the reservoir and fate of the dam motivate some projects. Others are driven by scientific research questions but are also relevant to the current management challenges.

Because Searsville Dam and the lake are located on Stanford's Jasper Ridge Biological Preserve, a number of ongoing studies are investigating various aspects of ecosystem structure and health in environments affected by the lake. For example, Alan Launer of Stanford's Center for Conservation Biology is leading studies of the bullfrog population (an invasive exotic) around the lake and the aquatic ecology of San Francisquito Creek below Searsville Dam. Philippe Cohen and David Freyberg recently completed a study examining the opportunities for maintaining some open water habitat at Searsville Lake under various possible dam management scenarios—

removal on riparian forests of the Great Plains. Emily Stanley of the University of Wisconsin and her associates are emphasizing the ecological ramifications of removal of low-head structures in Wisconsin. Randy Parker of the USGS is conducting physical experiments with dam removal scenarios using flumes in a laboratory setting. Formal publications on these efforts are not yet available.

Perhaps the most extensive investigation of the effects of dam removal is being undertaken by Karen Bushaw-Newton at the Patrick Center for Environmental Research of the Academy of Natural Sciences in Philadelphia (Box 1.4). The research is focused on the removal of a

Box 1.3 continued

dam lowering, dam removal, and no alterations to the dam. Several additional research projects are likely to begin in the near future in connection with the assessment of alternatives.

In addition, the university, through consultants, has completed several management studies focused on both short- and long-term decision-making. Several studies have examined alternative ways to minimize flooding in the Family Farm Road area caused by sediment aggradations at the delta of Corte Madera and Sausal creeks. The university has implemented some engineering measures (i.e., channel improvements and hardening, culvert replacement, berm construction), and has developed an ongoing channel maintenance program based on these studies. Another recently completed study examined the seismic safety of the dam. An important study nearing completion examines the consequences of increased sediment delivery below Searsville Dam on downstream habitat and flood risk.

Broader-scale research in the entire watershed is being conducted under a coordinated resources management plan (CRMP), as well as a joint powers authority (JPA). The CRMP (involving more than 30 stakeholder groups, including local and state governmental agencies, major landowners, and local and regional community organizations) has completed a biological resources inventory for the riparian corridor that has been incorporated into a geographical information system. The JPA (members include the Santa Clara Valley Water District, San Mateo County Flood Control District, and cities of Palo Alto, Menlo Park, and East Palo Alto) is seeking federal funding for a U.S. Army Corps of Engineers reconnaissance study for a federally funded flood control project.

For more information, contact David Freyberg, associate professor, Department of Civil and Environmental Engineering, Stanford University, at freyberg@cive.stanford.edu.

small, run-of-the-river dam across Manatawny Creek, a tributary of the Schuylkill River whose waters eventually flow into Delaware Bay, in southeastern Pennsylvania. This multidisciplinary effort is scheduled for completion in 2002.

Although there are few refereed journal articles on the effects of dam removal, some of the scientific investigations and results are beginning to be reported in oral presentations at scientific meetings. These preliminary reports provide an indication of the likely sorts of information and ideas that will emerge from the ongoing research. In the roughly two years prior to the publication of the present report in early 2002, a variety

Box 1.4 An Ecological Approach to Small Dam Removal: The Manatawny Creek Study

In late 1999, the Patrick Center for Environmental Research and University of Delaware, in collaboration with the Greater Pottstown Watershed Alliance (GPWA) and Delaware Riverkeeper Network (DRN), began an integrative study of the effects of dam removal on lower Manatawny Creek in Pennsylvania. A low-head, run-of-river dam was removed from Manatawny Creek in a two-stage process, with the first half of the dam removed in August 2000 and the remainder removed in November 2000. The orphan dam was removed because the local watershed organization, GPWA, and DRN were concerned about the integrity of the Manatawny Creek system and raised the funds to pay for the removal. Funds for the ongoing scientific and restoration studies have been provided through grants from the Pennsylvania Department of Environmental Protection.

At sites upstream and downstream from the dam, scientists have been monitoring several physical, chemical, and biological components of the stream system to document the spatial and temporal changes associated with the removal. These components include geomorphology, sediment and water chemistry, algae, macro invertebrates, freshwater mussels, fish, and riparian vegetation. In conjunction with the removal, the DRN has begun restoring the riparian corridor in the former impoundment. This project represents one of the first comprehensive studies attempting to document the large-scale physical, chemical, and biological changes in a river system following dam

of scientific societies hosted special sessions focusing on the effects of dam removal. In August 1999, the American Fisheries Society hosted a special session on dam removal at its annual conference in Charlotte, North Carolina. The Geological Society of America hosted a special panel discussion at its national meetings in October 2000, at which participants tried to sort out the contributions of earth science to decision making regarding dam removal. In February 2001, the Association of American Geographers presented two sessions of papers and reports by researchers involved in dam removals. The topics emphasized the building of conceptual frameworks for the science. In August 2001, the Ecological Society of America hosted a special session on dam removal, which, like the sessions offered by the other organizations, extended the discussion beyond science to legal and policy dimensions. In June 2001, the North American Benthological Society hosted three special sessions on dam

Box 1.4 continued

removal. The results of this study will provide information to both scientific and management communities on the effectiveness of dam removal in the restoration of riverine ecosystems.

Visual documentation of the effects of dam removal on the Manatawny Creek system before, immediately following, and one year after removal.

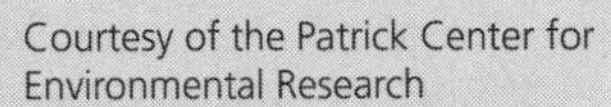
Courtesy of the Patrick Center for Environmental Research

For more information on this project, see http://www.acnatsci.org/research/pcer/manatawny/.

removal that included wide-ranging assessments of the ecological results of removals.

Despite the early stage of development of the science of dam removal, structures *are* being removed, and short courses are beginning to appear to provide advice for decision makers, engineers, and scientists involved in the process. Exemplified by an annual course offered at the University of Wisconsin, Madison, these short courses show that the engineering community, in particular, is beginning to build a level of experience with the process of dam removal, and that there is considerable experience with the decision-making and political processes of removal.

An unfortunate disconnection occurs between the research pursued by the academic community and the research needed by decision makers. The majority of published research on the downstream impacts of dams, research that is likely to be informative about the potential outcomes of dam removals, focuses on the effects of large and very large dams. The effects of these dams are more easily seen, more obvious, and more easily mapped or measured than is the case with smaller dams. For example, after the expenditure of about $100 million by the federal government, more scientific literature, data, and understanding exists for the Colorado River downstream from Glen Canyon Dam than for any other location. However, at present and for the near future, decision makers are most concerned with small and (to a lesser degree) medium-sized structures. There is much less research available on the effects of these dams on physical and biological components of ecosystems. As a result, decisions must be made with relatively little scientific support. As shown by the preceding review of ongoing research on the outcomes of dam removal, the disconnection between scientific research supply and demand has begun to be rectified, but the gap between available and needed theory and knowledge is still substantial.

Another disconnection is apparent within the research base itself. High-quality research began to appear in 2001 and 2002 addressing questions about the effects of dam removal, but many projects are moving forward in isolation from similar work elsewhere. The questions and approaches that are most important to geologists may be very different from the questions and approaches significant to ecologists, whereas the issues faced by planners, legal experts, property owners, and dam owners call for answers to still other challenges.

The most pressing need is for much-improved integration of scientific efforts. Although groups of scientists from each discipline are beginning to present their work to colleagues working in the same discipline, there is little evidence of a truly broad-gauged, multi-science dialogue, despite the fact that the study of effects of dam removal is highly integrative. Decision makers in the public arena need to deal with the entirety of the effects of dam removal to reasonably assess a variety of trade-offs. Unless the various sciences are able to work across intellectual boundaries, their contributions to decision making will be diminished.

CONCLUSIONS AND RECOMMENDATIONS

Dams are a ubiquitous feature of the American landscape and waterscape, and they form an integral part of the nation's economic infrastructure. The building of many of these structures has produced significant economic benefits, but the effort also has imposed environmental, economic, and social costs that are now becoming clear. The majority of structures are small, storing less than 100 ac ft of water, and most small dams in the nation are owned by private concerns or local entities. An unknown number of dams have been removed, but the total is probably at least 1,000. The removal of these structures, mostly small, run-of-river dams, typically has been the result of decisions by individual owners seeking a variety of largely economic benefits, although the environmental reasons for dam removal are numerous and often supported by local or state governments.

- **Conclusion:** Science to support decisions about dam removal is progressing, but there is little cross-disciplinary communication, and research priorities have not been established to guide researchers or funding efforts.
- **Recommendation:** The panel recommends that federal agencies and other organizations consider sponsoring a conference for researchers who are focusing on the scientific aspects of dam removal with the specific objectives of improving communication across disciplinary boundaries, identifying gaps in knowledge, and prioritizing research needs. The conference should not be a forum for debating whether dams should be removed, because other venues are available for bringing stakeholders together. The conference should focus on science and the state of knowledge available for decisionmakers, identify gaps, and assign priorities.

- **Conclusion:** Dam removal is a site-specific process, largely dependent on the owner and often in collaboration with local stakeholders and state and local government. These decision makers need more information and a framework for effective decision making. Data about dams that have been removed can be useful for decision makers considering the fate of existing structures, yet there is no centralized mechanism for collecting, archiving, and making available such information on a continually updated basis.
- **Recommendation:** When dams are removed, their entries in the National Inventory of Dams are deleted and the National Performance of Dam Program retains information about them. The panel recommends that federal agencies improve the availability of information about dam removal by making this database widely known and available to the public.

2

The Federal Legal Context Affecting Dam Removals

Federal, state, and local laws and regulations at every step of the decision-making process influence dam removal decisions. In some instances, a law or regulation may stimulate the debate in the first place, such as when a dam removal is required to restore upstream habitat for a species of fish listed under the federal Endangered Species Act (ESA). The possible involvement of a wide range of laws and regulations needs to be considered when the decision to remove a dam is being made, as well as during the actual removal. For instance, the federal Clean Water Act (CWA) may apply if the dam's removal and release of sediment from the former impoundment would change pollutant-loading levels or affect temperatures downstream. This chapter focuses on the federal policy context of dam removal; it is outside the scope of this report to examine the context in all the states, which differ greatly in terms of the details of their laws, policies, and programs that may affect dam removal. In general, the removal of federally owned dams is governed by federal agencies and subject to the availability of appropriated funds. In contrast, the removal of privately owned dams are governed primarily by state and local rules, although many federal laws and regulations still could be relevant. State governments empower local governments to engage in land- and water-use planning, zoning, and taxation, and most states delegate authority to local governments to regulate subdivisions and provide local public infrastructure. Dam owners and others who may be evaluating possible dam removal need to seek guidance from their local jurisdiction and the state agency with jurisdiction.

There is no comprehensive, consistent national policy on the removal of dams, nor are there specific federal regulations or policies governing dam removal. However, the federal government plays various roles in the context of a dam removal. Many federal agencies can exercise jurisdiction in such a decision, including the Federal Energy Regulatory Commission (FERC), U.S. Department of the Interior, U.S. Army Corps of Engineers (USACE), U.S. Environmental Protection Agency (USEPA), and U.S. Department of Agriculture (USDA). In addition, numerous federal statutes and programs are relevant to the construction, alteration, and operation of dams and could be relevant to dam removal. The most important of these are the CWA, ESA, and National Environmental Policy Act (NEPA), all mentioned in the previous chapter; as well as the Federal Power Act of 1920 (P.L.16 USC 791a), Electric Consumers Protection Act (ECPA) of 1986 (P.L. 99-495), National Historic Preservation Act (NHPA) of 1966 (P.L. 89-665), western water rights law, Small Watershed Rehabilitation Amendments of 2000 (P.L. 106-472), Indian Dam Safety Act of 1994 (P.L. 103-302), National Dam Safety Program, and FERC Dam Safety Program.

HYDROELECTRIC DAMS

The potential removal of any private, municipal, or state hydroelectric dams involves FERC, an independent regulatory agency that licenses and inspects hydroelectric projects. Federal hydroelectric dams, in contrast, are authorized by Congress and constructed by the U.S. Bureau of Reclamation, USACE, or Tennessee Valley Authority and subject to NEPA requirements.

An Act of Congress created FERC in 1977. At that time, FERC's predecessor, the Federal Power Commission, was abolished and FERC inherited most of the responsibilities that were first granted in the Federal Power Act of 1920.* The owners of hydroelectric dams must reapply to FERC for an operating license every 30 to 50 years. In the relicensing process, the dam owner must show that the dam operation continues to be in the public interest. Since 1986, when the Congress passed the Electric Consumers Protection Act (ECPA), FERC has been required to give the same level of consideration to nonpower values (e.g., the environment, recreation, fish and wildlife) that it gives to power and development

* The Federal Power Act can be reviewed online at http://www.ferc.fed.us/intro/acts/fpa.htm.

Box 2.1 The Removal of Edwards Dam on the Kennebec River in Maine

The 1999 removal of Edwards Dam in Maine marked the first time that the Federal Energy Regulatory Commission (FERC) ever ordered a dam to be removed against the wishes of its owner (American Rivers et al., 1999). In 1993, the 30-year license to operate Edwards Dam expired. The Edwards Manufacturing Company and city of Augusta, the dam's owners, applied to FERC for a new 30-year license that would allow them to continue to operate the 150-year-old hydroelectric dam. They also asked permission to increase the amount of electricity generated by the dam from 3.5 megawatts to 11 megawatts (American Rivers, 2001b). In return, the owners would install interim fish passage facilities to offset the environmental harm caused by the dam while FERC prepared its Environmental Impact Statement (EIS) as part of the relicensing process (American Rivers, 2001b). The National Environmental Policy Act requires an EIS. Since the passage of the Electric Consumers Protection Act in 1986, FERC has had to give the same level of consideration to non-power values (e.g., the environment, recreation, fish and wildlife) that it gives to power and development objectives.

In January 1996, FERC released its draft EIS recommending that the commissioners relicense the dam and require the owners to construct fish passage for seven target species. The Kennebec Coalition, a group of four nonprofit environmental groups that formed in 1993 when the original license expired, filed extensive comments with FERC, claiming that dam removal should have been chosen as the preferred alternative for both biological and economic reasons (American Rivers, 2001b). The FERC staff took a second look at the

objectives when deciding whether or not to relicense a project. The ECPA also increased opportunities for agencies, interested organizations, and the public to participate in the process and required FERC to base its recommendations for mitigating adverse effects of a licensing proposal on the recommendations of federal and state fish and wildlife agencies and to negotiate with the agencies if disagreements occur.

In 1992, the Congress further altered FERC's hydropower program under the National Energy Policy Act (P.L. 102-486). The Act prohibits licensees from using the right of eminent domain in parks, recreational areas, or wildlife refuges established under state law. It allows applicants for licenses to fund environmental impact statements—referred to as third-party contracting—and authorizes the commission to

Box 2.1 continued

issues, and when the final EIS was released in July 1997, they recommended that the commissioners deny the relicensing application and order the removal of the dam (Costenbader, 1998). The studies required by the EIS found that removing the dam would allow nine species of migratory fish access to 15 miles of historical spawning habitat and result in an overall increase in wetland habitat, recreational boating, and fishing opportunities; that installing fish passage facilities (required for the relicensing) would cost 1.7 times more than removal; and that customers paid twice as much for the dam's electricity as they would have for other sources of power (e.g., combustion turbine plants or natural gas) in the region (Federal Energy Regulatory Commission, 1997).

On November 25, 1997, the five FERC commissioners voted to deny the relicensing application and ordered Edwards Dam to be removed at the owners' expense. Subsequently, the owners filed a request with FERC for a new hearing and a stay of the order, which FERC granted (American Rivers, 2001b). The owners agreed to remove the dam but argued that they should not have to bear the full cost of removal. In May 1998, to avoid a lengthy court battle, all parties actively involved in the relicensing process signed a settlement that transferred the dam's ownership to the state of Maine. The costs of removal and fish restoration efforts were borne by upriver dam owners in exchange for a delay in their fish passage obligations, and a downriver shipbuilder in exchange for a permit to expand its shipyard operations (American Rivers et al., 1999). In 1999, Edwards Dam was removed. Currently, FERC's authority to order the removal of Edwards Dam is being debated in court because the cost of removal was not borne by the owner but rather by stakeholders who gave funds as in-kind mitigation for impacts at other water facilities in the area.

assess licensees for costs incurred by fish and wildlife agencies and other natural and cultural resource agencies for studies required under Part I of the Federal Power Act.

None of FERC's enabling legislation sets forth procedures for removing hydropower dams. In the absence of specific rules covering removal, one would expect relevant agencies to review a removal plan as they would a dam construction or modification program. FERC and other relevant agencies can be expected to develop appropriate policies as removal proposals are implemented. FERC also has the power to deny the relicensing of a hydroelectric dam, an action that could result in a dam being removed (Box 2.1). Typically, if FERC denies an application for a license renewal, another party can claim the license, and whatever problems led to the license denial may be mitigated. If no one claims the license, then the dam may be removed.

DAM SAFETY PROGRAMS

The average life expectancy of a dam is 50 years, and a full 25 percent of all U.S. dams are now more than 50 years old. By 2020, that figure will reach 80 percent (American Rivers et al., 1999). Often, safety is a factor in a decision about whether or not to remove a dam. As dams age, concern over their safety grows, and oversight and a regular inspection program are extremely important. Moreover, the issue of potential removal of these older dams is likely to become more significant in the future. Many agencies are involved: FERC, FEMA, USACE, Bureau of Reclamation, Tennessee Valley Authority, Bureau of Land Management, Fish and Wildlife Service, National Resource Conservation Service, and Bureau of Indian Affairs all administer dam safety programs at the federal level.

National Dam Safety Program

Section 215 of the Water Resources Development Act (WRDA) of 1996 (P.L. 104-303) established a National Dam Safety Program under the jurisdiction of FEMA. The purpose of the program is to "reduce the risks to life and property from dam failure in the United States through the establishment and maintenance of an effective national dam safety program to bring together the expertise and resources of the federal and nonfederal communities in achieving national dam safety hazard reduction" (WRDA §215[a]). The National Dam Safety Program does not specifically govern or regulate dam removal. It is relevant, however, in addressing a variety of actions that modify dams.

The law requires FEMA to establish an Interagency Committee on Dam Safety, which FEMA now chairs, and a National Dam Safety Review Board. It also requires FEMA to coordinate federal dam safety efforts in cooperation with state dam safety officials, transfer knowledge and technical information among federal and nonfederal agencies, and provide for public education in the hazards of dam failure and related matters. FEMA also is authorized to provide grants to states to establish and maintain dam safety programs and provide training for state dam safety staff and inspectors. To the extent that safety issues encompass dam removals, FEMA addresses dam removal concerns under the National Dam Safety Program.

FERC Dam Safety Program

Dam safety is also an important part of FERC's hydropower program, although dam removal is not a topic on which the commission focuses its attention. In terms of number of dams inspected, the commission's dam safety program is the largest in the federal government. Of the approximately 2,600 hydroelectric dams that fall within FERC's domain, more than two-thirds are more than 50 years old.

Safety issues are present at every stage of a dam's life. Before dams are constructed, the FERC staff reviews and approves the designs, plans, and specifications of dams, powerhouses, and other structures. During construction, FERC staff engineers frequently inspect a dam. After construction is completed, FERC officials inspect the dam on a regular basis to verify the structural integrity, identify needed maintenance and remedial modifications, ensure proper maintenance, and verify that licensees comply with the terms and conditions of their licenses. Inspection visits are coordinated with resource agencies, state dam safety officials, and other interested agencies. The FERC staff also inspects dams on an unscheduled basis.

Every five years, an independent consulting engineer approved by FERC must inspect and evaluate dams higher than 32.8 feet (10 meters), or with a total storage capacity of more than 2,000 acre-feet (2.5 million cubic meters). The engineer identifies any actual or potential deficiencies that might endanger public safety and requires the dam owners to correct them.

The FERC staff also evaluates the effects of potential and actual seismic and large flood events on the safety of dams. The commission monitors and evaluates seismic research in geographical areas where there is concern over possible seismic activity. This information is applied in investigating and performing structural analyses of hydroelectric projects in these areas. During and following flood events, the staff visits dam sites; determines the extent of damage, if any; and directs any necessary studies or remedial measures that the licensee must undertake.

Lastly, FERC requires licensees to prepare emergency action plans and conducts training sessions on how to develop and test these plans.

The plans are designed to serve as an early warning system if there is a potential for, or sudden release of water from, a dam failure or accident involving the dam. The plans include operational procedures that may be used, such as reducing reservoir levels and reducing downstream flows, and procedures for notifying affected residents and agencies responsible for emergency management. These plans are updated and tested frequently.

Indian Dam Safety Act

The Indian Dam Safety Act of 1994 established a dam safety maintenance and repair program within the Bureau of Indian Affairs to maintain identified dams on Indian land that, if they failed, would present a threat to human life. Potential dam removals on tribal lands need to be evaluated in the context of this program.

PROTECTION OF NATURAL SYSTEMS

National Environmental Policy Act

The National Environmental Policy Act of 1969 is a general statute that declares a national environmental policy and promotes the consideration of environmental concerns by federal agencies. NEPA has had a pervasive effect on the federal decision-making process and has influenced thousands of projects and activities of federal agencies as well as state and local governmental and private projects involving federal funding or other significant federal involvement.

NEPA establishes national environmental policy and goals, provides a method for accomplishing those goals, and includes guidance on the fundamental question of how NEPA relates to other federal laws. NEPA announces a commitment to use all practicable means to conduct federal activities in a way that will promote the general welfare and be in harmony with the environment. NEPA's goals are intended to assure safe, healthful, productive, and aesthetically and culturally pleasing surroundings for all generations of Americans. Because NEPA creates no new substantive rights, its importance stems almost entirely from procedural provisions designed to ensure that federal agencies consider the environmental consequences before taking an action.

NEPA requires that an environmental impact statement (EIS) be issued for certain "major Federal actions significantly affecting the quality of the human environment" (NEPA §4332[2][c]). An EIS is a lengthy document based on an exhaustive process of public hearings, interagency consultation, and environmental research and analysis, including evaluation of alternatives and selection of a preferred course of action.

Compliance with NEPA matters very much to private and nonfederal government interests dependent on federal permit decisions, such as dredge-fill permits under Section 404 of the CWA. In addition, NEPA continues to be a primary basis for challenges to public or private land development decisions, most of which can be argued to have an environmental component. NEPA is important to environmentalists because it provides a statutory basis to force the review of federal decisions, regardless of whether the federal agency involved has distinct environmental responsibilities. The U.S. EPA also has a review role in NEPA under section 109 of the Clean Air Act.

NEPA created the Council on Environmental Quality (CEQ) as a part of the Executive Office of the President and defined its responsibilities. The CEQ has promulgated regulations that guide the NEPA process (*Code of Federal Regulations*, Title 40, Section 1500 et seq.). The CEQ is charged with monitoring progress toward achieving NEPA's national environmental goals and is required to assist and advise the President in the preparation of the environmental quality report. It is also the duty of the CEQ to gather environmental information and conduct studies on conditions and trends in environmental quality. In addition, the CEQ has been assigned the duty of providing guidance to other federal agencies on compliance with NEPA.

NEPA most likely will be a consideration in planning, designing, and carrying out a dam removal project, particularly if federal land is involved, federal funding is provided, or significant federal permits must be issued. Each of these factors could become a trigger establishing a "major federal action" requiring NEPA studies and reporting requirements.

For any proposal to remove a federal dam, NEPA applies, and in many cases, an EIS is required because of the significant alteration of the environment involved. When a proposed dam removal involves private or nonfederal government land and water resources, the NEPA process still could be triggered if there is sufficient federal involvement. For example, if federal funding is provided for the dam removal, or if the project cannot proceed without issuance of federal permits or other approvals, then NEPA applies.

Clean Water Act

The CWA is the principal law governing the quality of the nation's waterways. Its objective is to restore and maintain the chemical, physical and biological integrity of the nation's waterways. The Act has been amended numerous times and given a number of titles. The 1972 amendments (P.L. 92-500) gave the Act its current form.

Although there is no CWA provision or associated regulation that specifically addresses dam removal, federally approved standards and regulations promulgated under the Act could influence dam removal decisions. For example, if a dam removal changes pollutant-loading levels in rivers or streams, the U.S. Environmental Protection Agency's total maximum daily load (TMDL) requirements may apply. If temperatures change markedly, temperature standards may apply. The CWA also could become the basis for federal, state, or tribal involvement in dam removal. If dam removal, for instance, requires dredge and fill operations or destruction of wetlands in the reservoir, a permit from the USACE most likely would be required under Section 404 of the Act.

Endangered Species Act

The purpose of the ESA is to protect endangered and threatened species in the United States. It establishes a policy that all federal departments and agencies must seek to conserve endangered and threatened species and use their authorities in furtherance of the purposes of the Act.*

The ESA requires all federal agencies, in consultation with, and using the assistance of, the departments of Interior or Commerce, ensure that any actions authorized, funded, or carried out by them do not jeopardize the continued existence of any endangered or threatened species, or result in the destruction or adverse modification of habitat of such species that is determined by the secretary of Interior or Commerce to be critical, unless an exception has been granted by the Endangered Species Committee (ESA §1536[a][2]).

The Act identifies prohibited actions related to endangered species and prohibits all persons, including all federal, state, and local govern-

---

*The text of the ESA can be found online at http://www4.law.cornell.edu/uscode/unframed/16/ch35.html or http://endangered.fws.gov/whatwedo.html#General.

ments, from "taking" listed species of fish and wildlife (ESA §1538), except as specified under the provisions for exemptions (ESA §1539). The verb "take" is defined as harass, harm, pursue, hunt, shoot, wound, kill, trap, capture, or collect, or to attempt to engage in any such conduct. The U.S. Supreme Court has defined "take" to include many forms of habitat modification that threaten the continued existence of species. Provisions include civil penalties, criminal violations, enforcement, and citizen lawsuits. Additional guidelines for the protection of marine mammals are established in the Marine Mammal Protection Act of 1972 (16 U.S.C. §1361 *et seq.*). Consultation procedures are administered by the Fish and Wildlife Service within the Department of the Interior, and the National Marine Fisheries Service within the Department of Commerce.

The ESA can be a catalyst for dam removal decisions. When habitat for a listed endangered or threatened species is believed by some to be jeopardized by the presence of a dam, a movement to modify or remove the dam can gather momentum. For example, listings of Snake River salmon have led to the evaluation of dam breaching alternatives for the four lower Snake River hydroelectric dams. Similarly, the listing of steelhead trout has fueled interest in the removal of Rindge Dam in California's Malibu Creek watershed; the dam is believed to interfere with steelhead spawning.

OTHER LEGISLATION AFFECTING DAM REMOVALS

National Historic Preservation Act

With the NHPA, Congress established a lead role for the federal government in promoting historical preservation and fostering conditions under which modern society and prehistoric and historical resources can exist in harmony. An underlying motivation in passage of the Act was to transform the federal government's position of indifference to historical sites into a role assuming responsibility for stewardship of historical areas for future generations. In the case of Rindge Dam, built in 1925–1926, for example, members of the Rindge family who support preservation of the dam argue that its historical significance needs to be considered in the federal decision-making process.

The NHPA requires federal agencies to assess the impact of proposed projects on historically or culturally important sites, structures, or objects within the sites of proposed projects. It further requires federal agencies to assess all sites, buildings, and objects on a project site to determine if any qualify for inclusion in the National Register of Historic Places. The Act also establishes a procedure for archaeological activities and a system of civil and criminal penalties for unlawfully damaging or removing important artifacts.

The national register is an inventory of historical resources maintained by the National Park Service pursuant to the NHPA. The inventory includes buildings, structures, objects, sites, districts, and archeological resources. It also encompasses significant properties that have not yet been listed or formally determined to be eligible for listing.

Under the NHPA, agencies must establish preservation programs consistent with their missions and the effects of their activities on historical properties. Agencies also must consider historical properties and designate qualified federal preservation officers to coordinate their historical preservation activities.

The NHPA created the Advisory Council on Historic Preservation (ACHP), an independent federal agency, to advise the President and Congress on matters involving historical preservation. The advisory council is authorized to review and comment on all actions licensed by the federal government that will have an effect on properties listed in the national register or eligible for such listing. The Act requires that a federal agency involved in a proposed project or activity is responsible for consulting with the state historic preservation officer (an official appointed in each state or territory to administer NHPA programs) and the advisory council.

Federal actions and materials subject to historical protection consideration include, but are not limited to, construction, rehabilitation, and repair projects; demolition; licenses; permits; loans and loan guarantees; grants; and federal property transfers related to historical places. The agency sponsoring such an activity is obligated to seek comments from the advisory council.

The NHPA could be a consideration in a dam removal if the dam structures involved are found to have historical, prehistoric, or cultural importance; if valuable artifacts are found on the project site; or if the actions required to remove the dam may jeopardize historical, prehistoric, or cultural resources. Proposed federal actions involved in a dam removal that affects such artifacts must take into account the potential outcomes, and agencies must consult with the advisory council and document potential outcomes in environmental reports, such as a NEPA environ-

Box 2.2 Preserving the Tumwater Dam in Washington State

Tumwater Dam on the Wenatchee River in Washington State is a good example of a dam left in place for historical preservation reasons—even though it originally blocked salmon runs. This small dam, just 17 feet tall, produced electricity to power locomotives passing through a tunnel in the Cascade Mountains. Steam locomotion could not be used in the tunnel because of the smoke, and therefore electrical locomotives were needed. This little dam is historic for that reason; it is also located along a popular tourist highway, making a tour stop almost obligatory. After a debate over dam removal, officials decided to build a fish ladder and leave the dam in place. Now fish pass upstream, the historic structure is preserved, and tourists get a lesson in both fish and history.

mental assessment or EIS. If sites determined to be eligible for listing in the national register are to be disturbed during a proposed dam removal, then additional surveys, testing, and site characterization are likely to be required (Box 2.2).

Western Water Rights Law

In the western United States, the waters of a state are publicly owned. A state grants permission to use the water (e.g., a water right) but the holder of the right does not own the water. The water right is transferred with the property when the land is sold. A water right specifies a point of diversion, place of use, rate of withdrawal, total volume of water to be used, and season for the use. The rights to construct and remove a dam could be determined by these rights to ownership and use.

The original purpose of western water law was to resolve conflicts among users, not to protect the quantity or quality of water resources. According to the doctrine of "prior appropriation," the bedrock of western water law, the first person to take water from a stream for beneficial uses has priority over all subsequent users. The priority date determines who gets water when the quantity is restricted. Both appropriation and riparian rights have functioned to detach water from the watershed by promoting dams and diversions.

Currently, western water is over-appropriated, largely because states routinely grant water rights for more water than actually is found in a river or stream. The water is diverted for irrigation or municipal use, run through turbines, or stored in reservoirs.

In 1992, the Congress passed the Western Water Policy Review Act (P.L. 102-575).* This Act authorized the U.S. Geological Survey and USACE to assist in a comprehensive review, in consultation with appropriate officials from the 19 western states, of the problems and potential solutions facing these states and the federal government in the increasing competition for the scarce water resources in the region. The Act authorized an advisory commission to

- Review present and anticipated water resource problems affecting the 19 western states
- Examine current and proposed federal programs affecting these states
- Review problems of rural communities relating to water supply, potable water treatment, and wastewater treatment
- Review the need and opportunities for additional storage or other means to augment existing water supplies, including, but not limited to, conservation
- Review the history, use, and effectiveness of various institutional arrangements to address problems of water allocation, water quality, planning, flood control and other aspects of water development and use
- Review the legal regime governing the development and use of water and the respective roles of both the federal government and the states over the allocation and use of water
- Review the activities, authorities, and responsibilities of the various federal agencies with direct water resources management responsibility

Tribal Governments and Water Rights

Tribal governments are considered sovereign governments under the U.S. Constitution. Tribal governments expect to participate as sovereign nations

* The text of this Act can be found online at http://www.den.doi.gov/wwprac/informat/act1.htm.

in dam removal decisions and discussions that affect tribal resources. In addition to the constitutional status accorded tribal governments, the federal government holds a "trust responsibility" for tribes. The trust is a product of Chief Justice John Marshall's commitment to recognize the indigenous nations' and tribes' inherent sovereignty within the context of a wider national government. In three decisions, he rationalized the federal government's power and held that the purpose of the exercise of the power was to fulfill the government's duty to protect the tribes' treaty rights (*Johnson v. McIntosh* 21 US, 8 Wheat., 543 [1823]; *Cherokee Nation v. Georgia* 30 US, 5 Pet., 1 [1831]; and *Worcester v. Georgia* 31 US, 6 Pet., 515 [1832]).

As applied to water, the trust responsibility requires that the federal government protect the tribes' continued enjoyment of their existing water rights. The Supreme Court's opinion in the 1908 case *Winters v. United States* (207 US 564 [1908]) remains the foundation of tribal water rights. At issue was the claim to the use of water from the Milk River in Montana by the Gros Ventre and Assiniboine tribes on the Fort Belknap Indian Reservation as against upstream non-Indian appropriators. The court recognized the "command of the lands and the waters" held by the tribes and the concession they had made to stay within the limits of the reservation, exchanging their nomadic life for a pastoral one.

Consequently, the extent of tribal claims to water resources is substantial. In 1984, the Western States Water Council estimated that tribal-reserved water rights might extend to as much as 45 million acre-feet. In most cases, tribal rights are senior to other water rights established under state laws.

In many treaties with the U.S. government, tribes did not give up their rights to water, but rather "reserved" the rights to continue fishing, hunting, and gathering in "all usual and accustomed places." These reserved fishing and hunting rights have been construed in several court cases to include an implied reservation of the water necessary to fulfill them (*United States v. Winters* [1908] and *United States v. Adir* 723F, 2d 1394, 1408-15 [9th Circuit (1983]). Moreover, because these reserved rights had been exercised since "time immemorial," the priority date of the implicitly reserved water right has been interpreted to extend back beyond the reach of memory, record, or tradition.

Further, the U.S. Supreme Court has ruled that when the federal government created the Indian reservations, it implicitly reserved the amount of water necessary to support present and future homelands. This is true whether the reservation was created by treaty or executive order.

The priority date of these implied water rights is the date of the reservation (Cohen, 1982; Pisani, 1996).

Native fish species that are to be protected under the tribes' reserved fishing rights include both anadromous fish, such as salmon and sturgeon, and resident fish, such as trout, whitefish, and sucker. Because these species have different life cycles, their needs vary, too. The natural river system provided a wide range of habitats that supported the native fish. Many native plant species that are culturally important to the tribes for food, medicine, or other purposes also have water needs, especially if they are adapted to riparian areas or marshes.

SMALL WATERSHEDS REHABILITATION AMENDMENTS

To address concerns over the safety of small flood control dams built by local communities with federal assistance, the Congress recently amended the Watershed Protection and Flood Prevention Act. The Small Watersheds Rehabilitation Amendments of 2000 (P.L. 106-541 Section 313) authorizes $90 million over five years to the Natural Resources Conservation Service (NRCS) to provide financial and technical assistance to organizations to cover a portion of the costs incurred for the rehabilitation of structural measures originally constructed as part of a covered water resources project. The FY 2002 Agricultural Appropriations Bill appropriated $10 million to begin implementation of the program. Over the past two fiscal years, Congress appropriated $16 million for pilot projects in four states. The NRCS will work with local community leaders and watershed project sponsors to address public health and safety concerns and environmental impacts of aging dams. The activities may include removing a structure if the sponsoring local organization requests it.

The amount of federal funds available is limited to 65 percent of the total rehabilitation costs and cannot exceed 100 percent of the actual construction costs incurred; none of this financial assistance can be used to perform operation and maintenance activities. The rehabilitation of structural measures must meet standards established by the Secretary of Agriculture and address other dam safety issues. The secretary also may provide technical assistance to a requesting organization in planning, designing, and implementing the rehabilitation projects.*

* For more information, see http://www.ftw.nrcs.usda.gov/pl566/agingwater/infra. html.

Wild and Scenic Rivers Act

The Wild and Scenic Rivers Act of 1968 (P.L. 90-542) protects U.S. rivers and their local environments that have remarkable scenic, recreational, geologic, fish and wildlife, historic, cultural, or other similar values. The Act preserves them in a free-flowing condition and designates three classes: wild, scenic, and recreational. Wild rivers have no development and are the river equivalent of wilderness. Scenic rivers have some evidence of human activities and some access points along their lengths. Recreational rivers have numerous pieces of evidence of human activities and many access points, and they may have undergone some impoundment or diversion in the past (National Park Service, 2001). The Wild and Scenic Rivers Act could be used to protect free-flowing rivers with significant natural and cultural resources. Rivers and streams included or proposed for inclusion into the system must be considered during project planning, and project impacts must be identified in an EIS. If a dam is removed from the river, the river is eligible for inclusion as a recreational river under the Act. At present, approximately 10,500 miles of rivers are included in the national system (National Park Service, 2001). Many states also have such designations, with substantially more river miles included.

3

Dam Removal Decisions

A DETERMINATION to keep or remove a dam needs to account for complex social, economic, and environmental interactions. This chapter presents a framework of steps and indicators to help decision makers assemble data and analyses that will assist them with this often difficult and contentious determination. The framework will help clarify multiple goals and objectives, identify the issues of greatest concern to a variety of stakeholders, and structure data collection and analyses. For owners or communities that decide to remove a dam, the chapter includes a discussion of post-removal monitoring and adaptive management to help them ensure that the original goals are achieved. The chapter begins with an introduction to the method used over the last 50 years to decide whether to build dams: benefit–cost analysis.

THE ECONOMICS OF DAM REMOVAL

Every existing dam was originally constructed for some explicit economic purpose. Many of the earliest small dams in the United States provided waterpower to mills and other industrial facilities. Some of these dams later were adapted to low-head hydroelectric generation, and many new hydroelectric dams were constructed. Medium-sized dams often were designed to provide a reliable supply of water for urban and/or agricultural uses. Other dams were constructed to provide storage for flood peaks, thus reducing the hazards to human life and flood-related property damage. Farmers have long used small dams to store water for livestock watering and fire suppression. By the last third of the twentieth century, property values in suburban housing developments frequently were enhanced by the creation of artificial "real estate lakes."

Because each dam had a known cost and economic purpose at the time of its construction, it is reasonable to assume that some type of financial or economic analysis justified each. For older dams, this justification likely was limited to a simple, nondiscounted comparison of construction cost to anticipated economic benefit. Only the costs and benefits affecting the dam owner would have been considered. Before the mid-twentieth century, dams were single-purpose projects. Even where multiple services were provided (e.g., a hydroelectric dam that also reduced downstream flood discharge), the dam was likely to be justified based on the primary purpose. The external costs of construction, including environmental costs, typically were omitted from the economic analysis or ignored entirely.

In 1936, the Congress authorized the U.S. Army Corps of Engineers to construct dams and other improvements for reducing flood hazards. The Flood Control Act (FCA) of 1936 (P.L. 74-738) contained a highly significant provision, little noticed at the time, stating that the policy of the United States was to construct improvements for the purpose of flood control where the "benefits to whomsoever they may accrue are in excess of the estimated costs" (FCA ch. 688, §1,49 Stat. 1570). This statement set into motion an investigation of the economics of dam building that culminated in the 1950 publication of the first formal procedures for benefit–cost analysis of water resource projects (U.S. Federal Inter-Agency River Basin Committee, 1950). These procedures were revised, elaborated on, and redesigned a number of times in subsequent decades.

Beginning as informal guidance, the benefit–cost procedures became formal guidance for all federal agencies involved in water resource development and ultimately were published as federal regulations in the late 1970s. In 1983, the regulations were rescinded and the procedures returned to their original status as informal guidance, known as the Principles and Guidelines. Over the years, however, the existence of written procedures or guidance had a profound effect on the design and evaluation of large federal dams, and, eventually, on the accepted framework for the evaluation of most dam proposals, both within and outside of government.

In particular, benefit–cost analysis practice came to incorporate a number of features:

- Even when motivated by a single issue, dams normally are designed and analyzed as multipurpose projects.

- For a dam to be economically feasible, the expected present value of the beneficial effects needs to exceed the expected present value of the adverse effects.
- Beneficial effects are to include the incremental value of all goods and services produced by the project as well as any beneficial external effects.
- Adverse effects are to include the costs of planning, designing, constructing, and operating the dam and reservoir, as well as any adverse external effects.
- The beneficial effects of each project's purpose are expected to exceed the allocated adverse effects of that purpose; otherwise, the project needs to be reformulated to eliminate that purpose.
- For each project, alternative structural and nonstructural solutions are to be defined and analyzed, so that the most economically efficient strategy is selected.

It is the nature of water resource development that the beneficial effects derive mostly from the known purposes of the project. Accordingly, they tend to be economic goods and services (e.g., water supply, electric energy, flood protection), which can be predicted and evaluated in monetary terms. Adverse effects also include some easily quantified components (e.g., construction and operating costs). But every water resource project has potentially large adverse external effects, such as the loss of habitat, fish passage disruption, destruction of wetlands, loss of wild river recreational opportunities, population movements, traffic congestion, and so on.

The stereotypical analysis produces three types of impacts: monetized benefits, monetized construction and operating costs, and nonmonetized (possibly nonquantitative) external costs. Not surprisingly, benefit–cost analyses of dam projects have been widely criticized for focusing on the quantitative effects while ignoring or minimizing nonquantitative external costs.

Additional considerations in assessing the financial aspects of a possible dam removal are the questions of who will benefit and who will pay. The distribution of costs and benefits among private individuals or companies and the public taxpayer is a pivotal issue for some removal decisions. An effective decision-making process exposes the true sources of financing and true sinks of additional costs before the decision is made. There is also a geographical aspect to costs and benefits; those who bear the costs seldom are in the same location as those who benefit. The entire general community may benefit from a dam removal, whereas most of the cost may be born by the owner of the dam and the owners of property on the shore of the reservoir that will be eliminated. Alternatively, the general community may pay the bill for dam removal, but the benefits may be enjoyed only by a dam owner who escapes financial and legal responsibility for the structure, and by property owners near the dam site that is eventually changed in some way by the removal.

INFORMED DECISION MAKING

Credible dam removal decisions take into account administrative, political, social, and environmental issues as well as factors emphasized in economic analyses. An informative general review of decision-making processes for dam removal is provided in *Dam Removal: A Citizen's Guide to Restoring Rivers* (River Alliance of Wisconsin and Trout Unlimited, 2000). However, much of this document focuses on how to remove a dam after the decision to remove it has been made. This section presents general guidance to help decision makers approach a decision to keep or remove a dam.

The process (Figure 3.1) begins with a clear identification and definition of goals and objectives. The articulation of these goals and objectives provides the framework within which the advisability of dam removal can be evaluated. Ideally, the procedure allows decision makers to compare the ecological, economic, and social outcomes of keeping or removing the structure. In addition, if a decision is made to remove the structure, the process will provide a foundation for continued monitoring and management corrections to ensure that the objectives are achieved.

This general method for reaching decisions about dam removal involves four basic steps:

Step 1: Define the goals and objectives

Step 2: Identify major issues of concern

Step 3: Data collection and assessment

Step 4: Decision making

If a decision is reached to remove the dam, two more steps are added:

Step 5: Dam removal

Step 6: Data collection, assessment, and monitoring

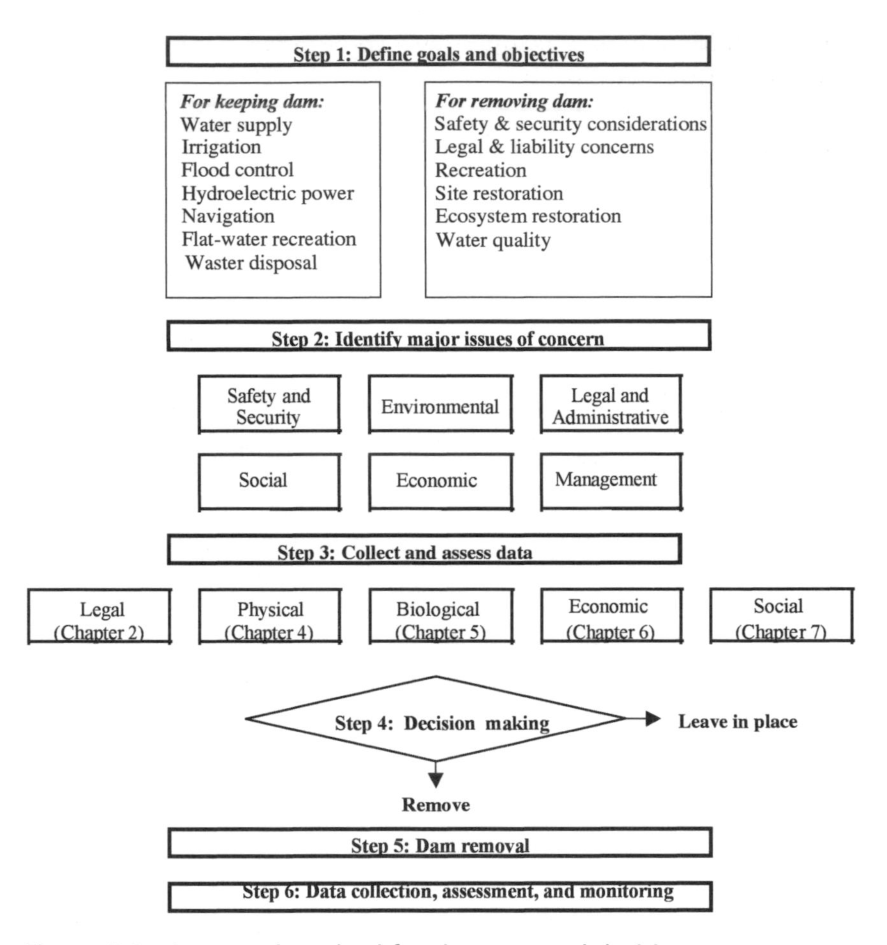

Figure 3.1 A general method for dam removal decisions

Step 1: Define the Goals and Objectives

To establish a basis for a dam removal decision, the goals and objectives for either removing a dam or leaving it in place need to be defined clearly. A stakeholder group needs to be assembled to help identify issues, concerns, and goals. Two key questions need to be addressed:

- Is the dam meeting its legally or socially defined original purpose and need?
- Have additional issues or needs arisen that need to be added to the list of goals?

Is the Dam Meeting Its Legally or Socially Defined Original Purpose and Need?

The first question challenges decision makers to evaluate the original purpose and need for the dam and determine whether the structure still meets its stated objectives. The typical reasons for dam construction, discussed in Chapter 1, are summarized below:

- **Recreation** is a significant by-product of many reservoirs created for other primary purposes. Flat-water recreation on reservoirs is common in the southeastern, midwestern, and Plains states.
- **Fire and farm ponds** are common in rural areas and are built primarily to impound water for livestock, agricultural, or fire-fighting uses. These ponds also serve as important recreational areas.
- **Flood control** is a major function of large, multipurpose dams in all parts of the nation, but especially in the East and Midwest. Medium-sized and large dams are used for flood control because large volumes of storage are required to capture potentially hazardous runoff and store it for subsequent gradual release.
- **Water supplies** for urban, domestic, and industrial use are obtained from systems made possible by dams. These dams range from small, run-of-river structures that divert stream flow into distribution systems to medium-sized and large structures that create reservoirs for temporary storage.
- **Irrigation** is made possible by dams. In the plains and western states, where rainfall is not consistent enough for the production of crops, low dams commonly are used to divert water for crop irrigation. Medium-sized and large dams create storage reservoirs in the upper portions of watersheds, filling them from runoff and snowmelt in winter and spring and releasing water for downstream diversions into lateral distribution systems during the growing season.
- **Waste disposal** is made possible by the construction and maintenance of dams that create holding ponds for use in several activities, particularly mining and industrial animal husbandry.
- **Waterpower** was the primary reason for the construction of many of the older dams in the United States. The advent of steam power made many of these structures obsolete for their original intended purpose, but many were refitted for other pur-

poses, including the production of electricity or have remained in place for aesthetic, cultural, or other reasons.

- **Electricity** is produced from dams of all sizes, ranging from low diversion works to large storage structures.
- **Navigation** on the nation's inland rivers depends on lock and dam systems that maintain pools of water deep enough to accommodate boat and barge traffic. Small dams raise water elevations, with boat and barge passage to and from different levels provided by locks adjacent to the dams. In upper reaches of watersheds, large storage dams impound reservoirs that release water to sustain the downstream pools.

Have Additional Issues or Needs Arisen That Need To Be Added to the List of Goals?

Societal preferences may have changed and additional objectives may have emerged since the dam was constructed. Accordingly, the second question requires decision makers to determine if these additional concerns have called into question the need for the dam. The typical reasons for dam removals, discussed in Chapter 1, are summarized below:

- **Dam safety and security** is a major issue in the consideration of possible removal. Dam failures can inundate downstream areas with unexpected floods and disastrous results. Dams installed 50 or 100 years ago at moderate cost now require substantial investments that are often several times the initial construction price simply to return the structures to safe, up-to-date operating condition. In many cases, if the owner is an individual or small business, removal is the only reasonable, economical alternative. In light of the potential for terrorist acts, security of dams and reservoirs also must be considered.
- **Liability concerns** can prompt action by dam owners, who may choose to remove a dam to eliminate their own potential liability. Turbulence downstream from small, run-of-river dams can be deadly traps for boaters and canoeists. People fishing from dams and related structures risk serious injury or drowning. The liability of a dam owner in the case of an injury or death is unclear, but some owners prefer to avoid the risk by removing the structure. The threat of liability for injury to life or property following a

dam collapse gives dam owners an economic incentive to repair unsafe dams or remove them, and removal may be cheaper than repair.

- **Recreation** can be used as a reason to support or oppose dam removal. Dams and their reservoirs make possible flat-water recreation, and dam removal, in addition to eliminating the reservoir, may enhance recreational opportunities downstream. For example, white-water boating in canyon rivers is enhanced by flows that are more natural. In flatland streams, canoeists and boaters seek continuous, uninterrupted lengths of river. Sport fishing, especially for trout in eastern and midwestern streams, benefits from rivers without subdivision by dams. However, some dams support trout fisheries that would not exist without the coldwater releases from reservoirs, and reservoirs often provide habitat for largemouth bass, a fish prized by anglers.
- **Site restoration** may be a benefit of a dam removal if the site where the dam is constructed is of historic, cultural, religious, or environmental importance. Often the dam site itself may be at issue, but the area below the waters of its reservoir also might be restored to support a newly designed ecosystem. The removal of dams also can contribute to the restoration of aquatic habitats downstream. Because dams artificially trap sediment and modify flow regimes, the reaches of rivers downstream are significantly altered from their original natural condition. The removal of dams may restore the movement of sediment in such systems as well as return the water to more natural temperatures.
- **Ecosystem restoration** is a possible benefit of dam removal. In addition to the obvious restoration of a river course inundated by reservoir waters, the river reaches downstream from a dam also can be restored to a more natural condition. The principal dam removal efforts to date have involved dams that fragmented streams and blocked spawning runs of anadromous fish, such as salmon and shad.

These questions address connections in the social, environmental, administrative, and political arenas. The definition of the overall goals and outcomes to be achieved by either retaining or removing the structure strengthens the decision-making process. After answering these questions, decision makers can (if there is sufficient justification) proceed to Step 2 of the evaluation process.

Step 2: Identify Major Issues of Concern

Once the goals have been established for either leaving a dam in place or removing it, the major controversies and specific issues of concern to various stakeholders need to be identified. This review needs to be accomplished in an open and transparent process, using the expertise and values of a wide array of people and institutions. The review needs to include the views of the owner of the dam and owners of land adjacent to the reservoir and along the downstream channel, as well as owners of water rights in the watershed. Local government agencies, along with state and federal regulatory agencies for water uses (consumptive and nonconsumptive), power, environmental quality, and fish and wildlife need to be part of the review as well. Nongovernmental organizations and groups advocating conservation, preservation, and economic development also are logical participants. Public sessions can provide a venue for input from individual citizens. The case of Rindge Dam (Box 3.1) is a useful example of the scientific, economic, and social complexity of dam decisions.

The widest possible involvement of stakeholders in the identification of issues is a key to success in reaching a sound decision about dam removal. At times, it may seem that an expeditious process is one that involves few participants. But agencies, organizations, and individuals with divergent opinions and viewpoints can reach compromise positions and innovative solutions most easily if they are all part of the decision-making process from the beginning, and if they are invited to participate rather than having to force their way into the process at a later stage. The early involvement of a wide range of participants also reveals potential problems when there is still ample time to address them, which is preferable to having the problems surface later in the process when there may be pressure to adhere to a schedule.

The types of issues that are likely to be raised, as illustrated in Figure 3.1, are safety and security, environmental, legal and administrative, social and economic, financial, and management issues. Clearly not all of these issues are contentious at any individual site, but the list below provides a good starting point for community discussions.

Safety and Security Issues

Identify safety and security issues associated with keeping or removing the existing structure. Questions to address include the following:

- Is there a significant potential for loss of life, injury, and property damage if the dam should fail or be removed?
- Is the dam vulnerable to failure because of either aging or inadequate maintenance?
- Is the dam vulnerable to acts of terrorism?

Environmental Issues

Identify environmental issues associated with keeping or removing the existing structure. Depending on the site, questions to address include the following:

- Will removal of the structure help to enhance the recovery of threatened or endangered species?
- Will removal of the structure lead to changes in unwanted invasive species or perhaps restore native species?
- Are there likely to be problems associated with contaminated sediments currently contained behind the dam if the dam is removed?
- Will removing the dam cause sediment to help build beaches?
- Will dam removal lead to a net gain or loss in wetland area?
- Have so many other changes occurred in addition to the dam that removal of the dam will not achieve the desired ecosystem restoration goals?
- What is the relationship of the dam and its removal to other parts of the watershed?
- How will drinking water supplies be affected?
- How will groundwater tables be affected?

Legal and Administrative Issues

Evaluate concerns and needs from a legal and process perspective. Questions that might be addressed include the following:

- Are there existing or potential conflicts with laws and regulations (e.g., Clean Water Act, Endangered Species Act) designed to protect natural systems?
- Are there existing or potential conflicts with laws and regulations (e.g., National Historic Preservation Act, tribal water rights) designed to protect social, historical, or cultural values?

Box 3.1 The Debate over Rindge Dam in Malibu, California

Rindge Dam in Malibu, California, was built in the 1920s by the Rindge family for use as an agricultural reservoir but was filled with sediment and nonfunctional by the 1950s. Today, in addition to no longer providing any water storage, the dam also blocks endangered steelhead trout from much of their habitat in the upper tributaries of Malibu Creek and contributes to erosion problems downstream on Malibu beaches. The dam currently is owned by the California Parks and Recreation Department and surrounded by state park lands. The decision-making process governing any future dam removal has been federalized because of the intervention of the U.S. Congress. A congressionally authorized study by the U.S. Army Corps of Engineers' Los Angeles District is under way to evaluate alternative means, including removal, of addressing the dam's presumed adverse impacts on the capability of the endangered steelhead trout to spawn in the upper reaches of the Malibu Creek watershed. The state of California is participating in this study.

Courtesy of Sarah Baish

The idea of removing Rindge Dam is not without controversy. Several homeowners in the Malibu Creek watershed whose homes are located in close proximity to the dam have expressed opposition to removal based on concerns about increased flood risk, and some members of the Rindge family oppose removal on historic preservation grounds. Malibu residents who have built expensive homes downstream of Rindge Dam in the Malibu Creek watershed, including Hollywood celebrities and some residents who have resorted to the courts for relief, apparently fear increased flood and mudslide risks because of removing the dam. On the other side of the debate, local conservation groups concerned about the survival of the steelhead support removal of Rindge Dam; these groups include California Trout, the Sierra Club's Malibu Group, American Rivers, and the Resource Conservation District of the Santa Monica Mountains. These groups, along with the U.S. Fish & Wildlife Service, California State Coastal Conservancy, Las Virgenes Municipal Water District, California Department of Fish & Game, National Marine Fisheries Service, and National Park Service, have created the Steelhead Recovery Task Force to investigate solutions and coordinate efforts.

- How does the existing structure fit into the overall management of the river system?
- Are there existing contracts for water supply and delivery?

Social Issues

Identify social issues associated with the existing project as well as those associated with removal. Examples of questions that might be addressed include the following:

- Are there changes in the types of, and access to, recreational opportunities?
- Are there effects on local and regional populations in terms of economic stability (or lack thereof), displacement, water supply, and loss of access to traditional use areas?
- Are there direct and indirect effects on the cultural relationships of the peoples to the landscape?
- Are there impacts related to changing regional and local economics?
- Are there direct and indirect impacts related to any necessary service that was provided by the dam, and how will this service be replaced?
- How will dam removal affect aesthetic property values in the area?

Economic Issues

Identify economic issues associated with the dam removal project. Examples of questions to be asked include the following:

- What is the cost of maintaining the dam versus the cost of other alternatives?
- Who is financially responsible for the dam and for any damage that might occur if the dam were breached? What are the potential costs (estimate) of any repair and annual maintenance of the existing facility?
- What is the status of the repayment on the debt for the project? Has it met the financial criteria defined in its authorization language?
- Are there financial criteria that must be met or maintained if the project is funded with international or public funds?
- Is the dam providing a service that will need to be replaced by some alternative, and what is its cost?

- What are the costs of alternative measures to mitigate for project impacts?
- What are the costs to provide additional security measures?
- How will property values be affected?

Management Issues

Identify the management issues associated with the dam and water control. Examples of questions to be addressed include the following:

- How does the existing structure fit into the overall management plan for the river system? Is it a critical element to meeting any legal agreements and providing a service to the local economy such as flood control, water supply, power production, irrigation, fire protection, or recreation?
- Do the operations fit into a broader context of river basin control?
- What is the source of funding for removal or restoration efforts?

From this series of questions, a suite of potentially contentious issues can be identified. This will help the decision makers and the public assess whether the dam should be considered for removal, what alternatives exist, and whether the process should move to Step 3, Data Collection and Assessment.

Step 3: Data Collection and Assessment

If, after the completion of Step 2, decision makers determine that there is reason and technical support to warrant further review, then data collecting and assessment need to be initiated.

The Heinz Center panel undertook two tasks to help decision makers better understand their choices. First, the panel developed a list of measurable indicators to support the decision-making process outlined in this chapter (Table 3.1). The dam owner, interest groups, scientists, engineers, and the public can use these indicators to gauge the potential outcomes of either keeping or removing an existing dam. To be of greatest use, such outcomes need to be forecast for various lengths of time into the future. The consideration of other rivers and streams that are similar to the one in question and can be used as points of reference, both with and without a dam, may be helpful in forecasting potential outcomes.

Second, the panel collected data and resources from scientific studies and previous dam removal projects that may be useful to decision makers. The remaining chapters of this report present information on the effects of dams, typical consequences of their removal, and, where possible, guidance for making site-specific forecasts of the consequences of dam removal. The qualitative descriptions and technical references included in these chapters are the best available resources for those seeking to gain an understanding of the consequences of decisions to remove or keep a dam.

If funding is available, The Heinz Center may prepare a handbook for communities offering more detailed guidance for site-specific evaluation. The Center would identify two communities that are currently considering the removal of a dam. A somewhat expanded panel would work jointly with community decisionmakers and the concerned public to identify key issues. With technical assistance from either a local university or state agency, the panel would assemble the relevant site-specific information using the indicators in Table 3.1. The goal would be to assist the two communities, but also to prepare a handbook for other communities to use.

Note that the present report does not include advice on evaluating dam safety. Handbooks on this aspect of the decision-making process are already available. Key sources of information include *Safety of Existing Dams* (National Research Council, 1983); *Safety of Dams: Flood and Earthquake Criteria* (National Research Council, 1985); and *Safety Evaluation of Existing Dams* (U.S. Bureau of Reclamation, 1980).

A fair amount of current and historical information is available from existing data collections, including some available free or at nominal cost from the World Wide Web. Web sources as well as traditional data sources are cited throughout each of the subsequent chapters. Geographic information (data displayed on maps) can be particularly helpful to decision makers; sources of geographic data and reliable base maps are listed in Appendix A.

STEP 4: DECISION MAKING

Once the data have been assembled, the scientific and economic assessments have been conducted with public input, and the legal review is completed, all the information needs to be forwarded to the ultimate

Table 3.1 Key Indicators for Making Dam Removal Decisions[a]

Potential Outcome Issue	Indicator
Physical	
River network segmentation	Length of free-flowing river
Watershed fragmentation	Percentage of watershed accessible to outlet of the river
Downstream hydrology	Flood frequency for bank-full discharge
	Measures for 100-year flood: magnitude, frequency, and duration
	Annual peak flow: magnitude, frequency, duration, and timing
	Diurnal flow variation
Downstream sediment system	Annual sediment yield
	Timing of maximum sediment yield
	Annual suspended load
	Annual bedload
	Mean particle size for bed and bank materials
Downstream channel geomorphology	Width of active channel
	Dominant channel pattern (single thread, braided, compound)
	Degree of channel sinuosity
	Frequency of islands, bars, beaches
	Spacing and frequency of pools and riffles or rapids
	Dominant channel process (aggradation or degradation)
Floodplain geomorphology	Degree of connection between floodplain and active channel
	Frequency of floodplain inundation
	Depth of floodplain inundation at various return intervals
	Areal extent of the annual and 100-year floodplain

(continued)

Table 3.1 *(Continued)*

Potential Outcome Issue	Indicator
Physical (continued)	
Reservoir geomorphology	Rate of sedimentation and sediment storage
	Rate of erosion and sediment loss
	Areal extent of delta wetland surface
	Length of shoreline
	Frequency and length of beaches, bluffs
Upstream geomorphology	Distance of upstream deposition or erosion
	Area subject to backwater flooding
Chemical	
Water quality	Turbidity
	Temperature
	PH (acidity or alkalinity)
	Levels of dissolved oxygen
	Concentrations of nutrients, toxins, heavy metals, radionuclides, herbicides, pesticides, and fuels
Sediment quality (reservoir area and downstream)	Organic content
	PH (acidity or alkalinity)
	Concentrations of nutrients, toxins, heavy metals, radionuclides, herbicides, pesticides, and fuels
Air quality	Pollution from boats
	Pollution from land-based vehicles
Ecological	
Aquatic ecosystems	Areal extent of aquatic ecosystems
	Productivity: primary, secondary, tertiary
	Diversity of species
Riparian ecosystems	Areal extent of riparian ecosystems
	Biomass of riparian vegetation
	Diversity of plant species
	Dominant plant species
	Number and extent of native, introduced, and endangered species

(continued)

Table 3.1 (*Continued*)

Potential Outcome Issue	Indicator
Ecological (continued)	
Fishes	Number and extent of native, introduced, and endangered species
Birds	Number and extent of native, introduced, and endangered species
	Connectivity and size of avian habitats
Terrestrial animals	Number and extent of native, introduced, and endangered species
Economic	
Dam-site economics	Income generated to the dam owner
	Relicensing costs
	Maintenance costs
	Operating costs and restrictions
	Required upgrading and refitting costs
	Removal costs
River reach	Value of urban/industrial water supply
	Value of irrigation water supply
	Value of navigation services
	Value of flood protection
	Value of hydropower production
	Value of waste disposal
	Local and regional recreation values for whitewater boating activities, flat-water boating activities, fishing, swimming, and shoreline recreation
	Property value gains or losses for reservoir shoreline and river banks downstream
Regional economic values	Number of jobs
	Value of water transportation or replacement
	Required additional investment for infrastructure: levees, channelization, bridges, locks, navigation equipment, canals, fish passage systems

(*continued*)

Table 3.1 *(Continued)*

Potential Outcome Issue	Indicator
Social	
Safety and security	Dam structural safety and security
	Potential for loss of life, injury, and property damage
	Public water supply vulnerability
	Vulnerability to failure from natural or human-induced causes
	Downstream implications of loss of dam
	Perceptions of safety of the reach
	Perceptions of safety of the reservoir
Aesthetic and cultural values	Aesthetic and historical values of the reservoir
	Aesthetic and historical values of the free-flowing river
	Religious values associated with the river and its landscape
	Historical value of the dam and associated structures
	Historical value of structures in and near the river
Non-majority considerations	Tribal sovereignty and rights
	Rights of minority populations, environmental justice
	Rights of future generations, intergenerational equity
	Animal and environmental rights

[a] Ideally, one would measure or estimate today's conditions and forecast conditions one year, five years, and a decade or more into the future

decision makers at the appropriate level. The ultimate decision whether or not to remove a dam is likely to balance the following concerns:

- Safety, security, and water management requirements
- Economics of maintaining the dam versus dam removal or other alternatives (i.e., alteration of the dam, change in operations)
- Ecological need and potential gains
- Societal considerations

- Legal relationships
- Public support and concerns
- Local, regional, and possibly national and international interests

If a decision is made to remove a dam, then the specific administrative process associated with complying with the National Environmental Policy Act (NEPA) or state equivalent may be needed.

Step 5: Dam Removal

The planning for and actual mechanical removal of dams is not covered in this report. Sources of information on the engineering aspects of dam removal are available elsewhere, including a periodic university short course on the subject. The Department of Engineering Professional Development at the University of Wisconsin, Madison, offers this short course, entitled "Succeeding with a Dam Decommissioning Project," at least once annually. The two-day course includes lectures and discussions on the engineering, social, economic, and environmental aspects of dam removal and is taught by professionals with experience in planning and engineering aspects of dam removal. The course includes an exchange of experiences from a wide range of states and can be a valuable lead-in to real-world dam removal cases for administrators, decision makers, and interested professionals.*

A good source of information on federal, state, local, and private funding mechanisms that can be used to finance dam removal and associated river restoration is *Paying for Dam Removal: A Guide to Selected Funding Sources*, issued by American Rivers (2000).

Step 6: Data Collection, Assessment, and Monitoring

Like many other projects involving rivers, dam removal projects require continuing management by responsible authorities, typically state-level managers. One approach to this process is adaptive management, which requires ongoing monitoring. The concept of adaptive management is essentially to learn by doing and adjust management strategies based on

---

*Additional information is available from the course director, Professor Patrick Eagan, by electronic mail at eagan@engr.wisc.edu.

the observed responses of a river to previous decisions. Its most effective form, active adaptive management, involves programs designed to compare selected policies or practices experimentally by evaluating alternative hypotheses about the system being managed (British Columbia Forestry Service, 2001). The process begins with specific management and technical decisions and actions based on predicted outcomes. Managers then scientifically evaluate the effects of these actions on a periodic, predefined basis through monitoring and measurement of a selected set of indicators. Then, based on the detected effects, management strategies are adjusted to ensure that the desired outcomes are achieved.

Adaptive management establishes a formal feedback process between management and monitoring so that decision makers can evaluate the effectiveness of their approaches. Adaptive management is not a license for endless research, but rather a method of taking the environmental, economic, and social pulse of the river on a periodic basis using low-cost monitoring of a few indicators. This set of indicators needs to be tailored to each specific case to ensure the collection of data that are useful from both scientific and administrative perspectives.

Monitoring is essential to evaluate whether the goals and objectives of dam removal are met. As stated earlier, any dam removal project needs to begin with the identification of project goals and objectives, whether to restore a natural ecosystem, improve safety, or raise property values. Goals need to be prioritized so that the project managers and evaluators have an understanding of their relative importance. Monitoring of the indicators defined in Table 3.1 after dam removal provides managers with a way to assess predicted outcomes. These data-based assessments provide repeatable examinations of goals and whether they are achieved.

Monitoring also can provide essential information to decision-makers and managers as they implement specific management activities. Currently, the concept of adaptive management is implemented by agencies in cases that call for complex or publicly sensitive decisions. Adaptive management is based on continual adjustments by managers in response to data collected by monitoring (Lee, 1993). If changes in indicators show undesirable trends, managers can make compensating adjustments. In adaptive management, there is a strong connection between science and management forged by a continual flow of data from monitoring (Wegner, 2000).

Monitoring and site-specific research are unlikely without commitments for sufficient staff and financial resources to handle the monitoring program. Therefore, monitoring needs to be identified directly and included in a dam removal budget. Data sharing and coordination of efforts among resource agencies, academia, and public and private researchers can reduce costs significantly.

CONCLUSIONS AND RECOMMENDATIONS

- **Conclusion:** Dam removal is a site-specific issue. The issue is complex because of competing values and competing regulatory issues, and therefore dam removal decisions require careful planning and review. To be effective and credible to managers, decision makers, and the public, a removal project needs to be informed by science, including social, economic, and environmental data. Sometimes the best available science is not enough, and additional investigations are needed. Decisions about dam removal take place in specific economic and social contexts that also need to be taken into account. Decision-making processes for dam removal are most effective when they are well organized, open, and inclusive of all the people in the affected communities.
- **Recommendation:** The panel recommends that participants in public decision making use a multistep process, beginning with the establishment of goals as a basis for the process, and including the identification of the full range of interests and concerns of those likely to be involved, the assessment of potential outcomes, and informed and open decision making.

    1. Identify the goals and objectives of the dam removal project.
    2. Identify the major issues of concern.
    3. Gather and assess the data.
    4. Decide whether to keep or remove the dam.

If a decision is made to remove a dam, then the following steps may apply:

5. Dam removal
6. Data Collection, Assessment, and Monitoring

- **Conclusion:** The assessment of potential outcomes of dam retention or removal requires measurable indicators that can be used to measure the present environmental, economic, and social conditions associated with the dam and to monitor future changes.

- **Recommendation:** The panel recommends that assessment of potential outcomes of a decision to retain or remove a dam include the evaluation of as many indicators as are applicable to the situation, with the assessments conducted for short-, medium-, and long-term periods, and for the "with dam" as well as "without dam" alternatives. The panel developed a list of measurable indicators (Table 3.1) that can be used to support the decision-making process.

4

Physical Outcomes of Dam Removal

Undisturbed rivers are in a state of partial equilibrium wherein the discharge of water, discharge of sediment, channel geometry, and geomorphic conditions are all in balance. Leopold et al. (1964) referred to this condition as "pseudo equilibrium" because rivers have so many forces acting upon them that perfect equilibrium conditions are rare. The installation of a dam on a stream introduces a new controlling factor, bringing about a new set of equilibrium conditions. In run-of-river dams, the new conditions may be very similar to the original pre-dam arrangements, except in the impounded reach, whereas if there is substantial water and sediment storage behind the dam, the new conditions may be quite different from those existing before the dam. In any case, given enough time, the river and dam establish a new balance of forms and processes.

A dam removal immediately introduces upstream and downstream changes to the river system. These physical, chemical, and biological changes are in part reversals of the outcomes that resulted from the dam's installation, and they represent adjustments of the river as it seeks an equilibrium with the conditions without the dam. This chapter discusses the background concept of physical integrity for rivers and the physical changes that are likely to occur as the result of a dam removal; these changes include reestablishment of fluvial dynamics in the impounded reach across space and time, reconnection of the segmented channel system, changes in hydrology, sediment dynamics, geomorphologic adjustments, and water quality changes. For each of these topics, the chapter reviews the effects of dams, outcomes of dam removal, measurable indicators of change, and sources of information.

PHYSICAL INTEGRITY*

The Clean Water Act (33 U.S.C. §§1251–1387) outlines the general policy for the nation regarding river and water quality. Section 1251(a) of the Act states that "The objective of this chapter [of law] is the restoration and maintenance of chemical, physical, and biological integrity of the Nation's waters." The Act contains specific actions that the federal government and others are to take to achieve this end. Although the Act does not define integrity, subsequent practice in the fields of water chemistry, hydrology, and biology have established some meanings. If a stream lacks chemical integrity, it is clear that its waters and sediments have chemical characteristics that pose health risks to humans and other organisms. In practice, the idea of chemical integrity is expressed in the assessment of chemical pollution and determinations of whether or not the chemical conditions of the stream water attain a defined chemical quality for its designated use, such as swimming, boating, or water supply (Sittig, 1980). Although the state of scientific knowledge is evolving with respect to exposure limits for many chemicals, the monitoring of the chemical characteristics of rivers is a straightforward technical issue.

Hydrologists and geomorphologists have conducted considerable research into the behavior of streams, but their investigations of the response of rivers to human interventions have been a relatively late addition to the mix of science for rivers (Brookes and Shields, 1996). The physical characteristics of the river environment are critical to understanding the chemical and particularly the biological components because the physical system is the stage upon which the chemical and biological systems are played out. The restoration of biodiversity to provide for a wide range of socially desired species, for example, depends first of all on the restoration of geo- and hydrodiversity, and it is precisely these components of the river system that are affected most directly by dams and, presumably, the removal of dams.

Much of what can be said or written about physical integrity rests on the collective experience of hydrologists and geomorphologists, who have spent most of their time investigating issues other than physical integrity. The concept is therefore undergoing change. Some states, including California and Arizona, under the sponsorship of the U.S. Environmental

---

* The ideas in this section are derived from research supported by a National Science Foundation grant to W. L. Graf.

Protection Agency, have formal statements regarding physical integrity that guide their river management decisions (e.g., Graf and Randall, 1998), and many state agencies are developing their own perspectives to fit their particular hydrologic, fluvial, and ecological conditions.

A broadly applicable statement defining physical integrity is the following:

> *Physical integrity* for rivers refers to a set of active fluvial processes and landforms wherein the channel, floodplains, sediments, and overall spatial configuration maintain a dynamic equilibrium, with adjustments not exceeding limits of change defined by societal values. Rivers possess physical integrity when their processes and forms maintain active connections with each other in the present hydrologic regime (Graf, 2001b).

In this concept of fluvial integrity, active processes driven by the flows of water and sediment are the keys to change, including the change initiated by the removal of a dam. Changes in flows stimulate changes in the geomorphology of a river, particularly the channels, floodplains, sediments temporarily stored in the system, and geographical or geometric characteristics of the river. Change is a continuing part of a river with physical integrity, but the reality of most American rivers is that change is limited by what society is willing to accept. From an economic standpoint, completely uncontrolled rivers are unlikely because of the need to protect financial investments associated with them. The present regime of a river refers to hydrologic processes that exist now rather than under some conceived set of conditions that once existed before there were any human impacts on the system.

SPATIAL AND TEMPORAL CONTEXTS

Now, the majority of dam removals take place as "targets of opportunity" in the sense that owners of individual dams begin the decision-making process in response to financial or safety issues. In many cases, orphan dams become candidates for removal because of their deteriorating condition, and states or local governments take over their ownership. An assessment of the potential outcomes of these individual structures is best undertaken in a watershed context. In the future, if larger numbers of dams become candidates for removal, or water resource and ecological values drive more decisions, a watershed framework would become essential in prioritizing candidates for removal to maximize restoration benefits.

Rivers are long-lived components of the earth's landscape. Some rivers, like the deceptively named New River of the central Appalachian Mountains, have existed for tens of millions of years. Truly "new" rivers are those that occupy the areas of the world recently abandoned by continental glaciers, such as northern North America and northwest Europe. In these landscapes, the rivers seen today have been active for a few thousand years. Even these relative newcomers, however, have been around longer than the technological effects that humans have imposed on them through dam construction. In the United States, technologically effective dams are mostly the products of the past two centuries. Therefore, a decision to remove a dam needs to take into account two time scales: a short one of a few decades, during which the river might reasonably be expected to change back toward its previous, undammed conditions (within constraints imposed by other controls, particularly land use by humans); and a long time scale, during which the river slowly adjusts to geologic and climatologic controls. These larger-scale forces form the backdrop for more immediate processes initiated by human decisions.

SEGMENTATION

Dams divide rivers into segments. Even without dams, rivers are segmented to some degree by changes in their hydrologic or geologic setting. For example, those places where major tributaries join a main channel, or where the channel crosses geologic faults or other structures, or where surface materials change substantially, all exert enough control to cause changes in river behavior. Dams are similar to these other natural dividers but are much more important because they physically divide the system and prevent the passage of sediment, alter the flow of water, and restrict the movement of organisms through the system.

The natural and human-created dividers along the lengths of rivers create a fluvial system that, although it is connected to a certain degree by the flow of water, is fragmented. The various divisions of rivers span a range of scales (Graf, 2001a):

- A *river* is the entire length of a stream from its formative point to the place where it empties into a body of water or larger trunk stream.

- A *segment* is a length of river that has as its beginning and ending points significant hydrologic or geologic boundaries, such as major tributaries, fault lines, or geologic structural or material changes. Segments are usually several tens of miles in length.
- A *reach* is a length of river with similar hydrologic, geomorphic, and ecological conditions throughout its extent. Reaches are usually one to a few miles long.
- A *site* is a cross section of river channel.

Generally, small dams affect sites and reaches of rivers, whereas medium-sized and large dams may affect segments and entire rivers downstream (Williams and Wolman, 1984). Physical changes resulting from small structures with little storage are likely to be fewer than the more far-reaching changes resulting from larger structures with substantial storage that enables the manipulation of downstream discharge regimes. Fragmentation occurs along main trunk channels, but it is also a feature of streams with many dams on tributaries. In some river basins, changes in the main channel are the cumulative effect of the many smaller tributary streams that are regulated. The Connecticut River in New England is a primary example. The removal of dams reestablishes lost connections among river reaches and segments.

The most extreme form of stress on rivers, especially in the arid western United States, is the complete appropriation of water flowing in a channel, either by direct withdrawal or by pumping from the riparian zone. Only slightly less extreme is the conversion of reaches of free-flowing rivers into a series of lake-like impoundments by dams. The character of rivers also is drastically altered when the connections between the channels and riparian zone of floodplains are severed by channelization, levees, and regulation of the flood regime (see Plate I, on the inside front cover).

At present, there are no nationally available databases that specifically describe the segmented nature of river networks. Such a database might be constructed easily by combining two existing geospatial datasets, the National Stream Reach File and National Inventory of Dams (NID). The U.S. Environmental Protection Agency maintains the National Stream Reach File as a geographical information system product depicting stream courses with convenient dividing lines. The lines are often culturally related, such as roads, or are related to the junctions of tributaries joining the main streams. The purpose of the divisions is largely administrative and for accounting of the water quality in the streams. For the larger purpose of assessing the connectivity of river reaches, the file could be modified to establish divisions imposed by dams. The resulting segments of rivers then could be managed and assessed from an ecosystem perspective as separate geographical units of the stream system, and proposed dam removals could be assessed based on the degree of reconnection they might offer. The locations of existing dams could be added to the stream reach file from data in the NID, which includes the latitude and longitude of existing structures. Problems to be overcome in such an effort would include the large sizes of the files involved, the question of their compatibility, and the dubious reliability of much of the location data in the dam inventory (Graf, 1999).

In the absence of the dataset suggested in the previous paragraph, one measurable indicator of fragmentation is the total length of free-flowing streams without intervening dams in a watershed or river basin. Such measures can be made directly from topographic maps available from the U.S. Geological Survey. Paper maps can be ordered from the survey online (http://www.usgs.gov) and often are available for purchase in sporting goods, map, and outdoor recreation stores. Topographic maps are available in digital form online.* Other maps that are useful for investigating the connectivity of rivers and the distances between obstructing dams include published paper maps from the U.S. Forest Service and Bureau of Land Management. These maps cover portions of the nation where federally administered land is common. At a local scale, useful maps for distance measures along streams are available from almost all state departments of transportation. These transportation-oriented products are usually published at a county scale and show most significant watercourses.

HYDROLOGY

The removal of small, run-of-river dams is unlikely to alter the downstream hydrology of streams because such dams do not impound significant amounts of water. In dry land portions of the nation, however, small dams often serve as diversion works that guide the entire low flow of

* One source is http://www.terraserver.microsoft.com, a site that also includes aerial photography for most parts of the nation. Another useful source is http://www.topozone.com, which provides digital topographic maps at a variety of scales.

streams into receiving canals. If these diversion works become obsolete or if their removal addresses some other priority social goal, their removal reestablishes a flow of water to downstream areas on a continuous basis. The discharge in such restored streams may be small in magnitude, but its continuous nature has important implications for the hydrologic underpinnings of the aquatic and riparian ecosystems connected with the stream (Malanson, 1993). Recharging of the groundwater supply near the downstream channel is nearly a certainty because of the direct connection between stream flow and groundwater (Dingman, 1994). Flows carry some sediment in many systems, so the restoration of water flow also means increased mobility for sediments.

Dams that have some storage capacity have measurable effects on downstream hydrology. In the most general sense, the extent of their effects is related to their storage capacity relative to the normal flow of the river and the engineering characteristics of their outlet works. The ratio of the total storage of the reservoir behind the dam divided by the average annual yield of the river (the total volume of water that flows past a site in one year) expresses the size of the structure as measured in hydrologic terms (Graf, 1999). This is a dimensionless measure of the size of the dam relative to the stream, a significant issue because a small dam may have far-reaching impacts on a small river, whereas a structure of the same size may have much less significant effects on a large stream (Petts, 1984). The ratio of storage over yield ranges from very small numbers to more than 10 for some very large dams (indicating that the dam can store a volume of water equal to 10 times the amount expected to arrive in a single average year). Dams that have storage capacities that approach one year's water yield of the stream are likely to have large upstream reservoirs and substantial effects on downstream hydrology. The removal of such dams, therefore, also is likely to have far-reaching outcomes.

In addition to the relative size of the storage pool, the engineering characteristics of the outlet works for medium-sized dams also affect the degree of influence the dam exerts on downstream hydrology. If the outlet works for the dam are small relative to the mean annual discharge of the stream, the dam is likely to effect major changes in the behavior of the river downstream. Flood control dams are often of this type, because their major function is to contain large volumes of water and release the water slowly to protect downstream areas. However, many flood control dams have been required to enlarge their emergency spillways to avoid overtopping; during major floods, these spillways will release large amounts of water, causing extensive downstream flooding and property damage. Water supply structures also may have small release systems. Because these dams have very limited capability to release large quantities of water over short periods, they tend to bring about substantial changes in the hydrologic regime of the river they control. Hydroelectric dams, on the other hand, may be equipped with a large outlet capacity so that they can generate large amounts of electricity on demand. Although water flows through the penstocks and turbines of these dams, from a hydrologic perspective they sometimes may operate like run-of-river structures. Glines Canyon and Elwha dams on the Elwha River of Washington State are of this type. Because of their operating rules and large outlets, they do not substantially change downstream flows over the short term (Pohl, 1999).

Dams with significant storage capacity and the capability to control releases have greater effects on downstream hydrology than do run-of-river dams. Some of these downstream effects may be partly reversible if a dam is removed; they also may be reduced by altered operating rules. The most common downstream hydrologic effects are reduced peak flows, altered low flows, reduced range of discharges, altered timing of flows, and changes in ramping rates.*

Dams reduce peak flows as a flood control measure by storing the high volumes of inflow to their reservoirs and then releasing the flow gradually. This arrangement is a significant change from natural conditions, because most river channels in dry land areas are adjusted to their flood discharges, and most floodplains in humid regions are defined by the periodic high flows of the nearby channels. When dams lower the peak flows, they decrease the physical integrity of the downstream river because the channel-forming discharge is reduced, and floodplains are not as extensively connected to the river. The dynamic connections among the various parts of the cross-sectional landscape of the river—its channel, islands, bars, beaches, and floodplains—no longer operate as an integrated system. The removal of the dam restores the peak flows (an objective achieved in some cases by changing the operating rules of the dam), and thus returns dynamic connections among the various parts of the river landscape downstream.

---

* General information on the effects of dams on downstream hydrology, as well as references to detailed case studies, can be found in Collier et al. (1996), Dynesius and Nilsson (1994), Graf (1988, Chapter 7), Petts (1984, Chapter 2), and Williams and Wolman (1984).

Dams alter low flows, sometimes increasing them and in other cases decreasing them. Dams constructed primarily to store and supply water for irrigation purposes increase low flow, especially in the summer growing season, because the whole point is to deliver water during dry periods. Similar situations often develop for urban or industrial water supply structures. In these cases, a fairly uniform moderate discharge replaces highly variable and generally lower discharges that occurred under pre-dam conditions. These elevated flows form different channels and different aquatic habitats than existed before the dam, but the conditions are reversed if the dam is removed.

Dams also can reduce low flows, particularly during exceptional dry periods that do not coincide with downstream delivery contracts. In these cases, dam operators may close the outlet works temporarily to save water, resulting in the desiccation of downstream reaches and segments. As a result, physical integrity is compromised completely, and the river ceases almost all physical processes. As a physical basis for the ecosystem, it ceases to function as a river, but this condition may be completely reversible with the removal of the diversion dam.

Because dams alter both high and low flows, they alter the range of flows experienced in the river over the course of a typical year. The difference between the highest and lowest flow in a single year is important because it controls the extent of the active channel and floodplain system on the landscape. If the range between the highest and lowest flows is large, as is the case for many natural systems, a substantial width of the riverine landscape is activated each year. If the range is reduced by the placement of a dam, then the width of the active portion of the landscape is reduced, the channel shrinks, and the amount of active floodplain shrinks as well. The result is a fluvial system that is much smaller than the pre-dam system. Such shrinkage is especially common in systems west of the Appalachian Mountains. The Platte River of the middle Great Plains, for example, was miles wide before the construction of upstream dams, but its present channel and floodplain widths are so narrow that fish habitats and riparian forest areas have been damaged severely (Williams, 1978b).

The annual timing of peak and low flows is a characteristic of the operation of the physical system that has important consequences for reproduction in the biological components of the ecosystem. Fish in the channel and plants on the floodplain have evolved with internal annual clocks that maximize their reproductive success by timing certain activities to coincide with the annual flood, which usually occurs in spring.

Seed production among native tree species, for example, is usually at a maximum during the period when the annual peak flow occurs (Stromberg et al., 1991). When dams are installed, the schedule for the annual peak flow often changes, and, as a result, the production of new seedlings for native vegetation is reduced. Exotic vegetation with different reproductive mechanisms sometimes gains a competitive advantage as a result. The removal of the dam restores the natural timing of peak flows and is likely to favor native vegetation and native fishes.

Ramping rates are the rates of change from low to high flows and back to low flows again. In most river systems these changes are gradual, requiring a period of several days or weeks to go from low flow conditions to peak flows (Figures 4.1 and 4.2).* An exception to this generality is the small- to medium-scale, arid-region stream subject to flash flooding with a change from no flow to high flow in a period of an hour or so. The installation of dams with sufficient storage and outlet works can enable operators to alter the discharge quickly, with dramatic changes occurring within a short period, measured in minutes. These rapid ramping rates cause significant physical and ecological problems through accelerated erosion of banks, especially if the change is a rapid decrease in discharge. However, these rapid ramping rates are being increasingly moderated during relicensing to a closer approximation of natural ramping, thereby reducing fish stranding and erosion. Pore-water pressure in bank materials equalized at high flows suddenly is not supported by water in the channel, and bank collapse becomes common. Accelerated erosion of channel fringes from this process destroys islands, beaches, and floodplain edges with associated sediment loading in the channel. The removal of dams returns the system to a more natural arrangement with slow ramping rates and gradual change.

The most important data available for monitoring changes in stream flow are gaging records of daily discharge ("gaging" is the technical spelling of gauging, and installations that gauge stream flow are called stream gages). The U.S. Geological Survey (USGS) is the nation's custodian for water data based on the 6,600 stream gages that it operates. Since the first continuous gage was installed in 1885, more than 18,000 sites have been gaged for varying periods of time (Wahl et al., 1995). At

* These two figures show the same trends, but the first uses discharge data whereas the second uses water depth. Both are provided here because planners and analysts prefer discharge data and the general public may want depth information.

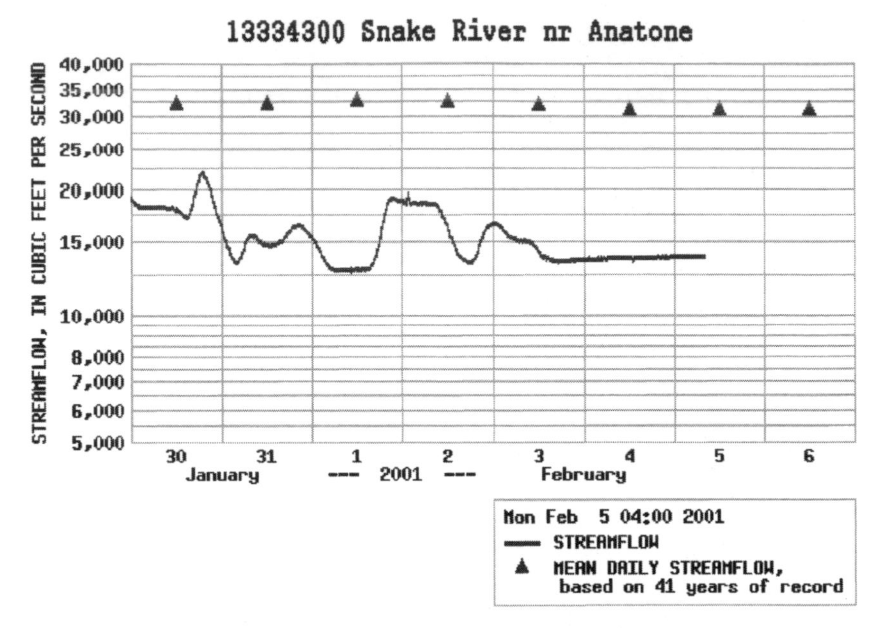

Figure 4.1 The operation of upstream hydroelectric dams creates steep ramping rates, as shown by this continuous stream flow record from the Snake River near Anatone, Idaho. See Figure 4.2 for a similar graph showing the changes in depth of flow that accompanied these discharge changes. *Source*: U.S. Geological Survey (http://www.usgs.gov/water).

present, about 7,000 sites are active. All of the data collected by the agency are available free of charge in digital form online (http://water.usgs.gov). The data are available for each day of record, as well as in an abbreviated form showing only annual peak flows. Information on each gaging station includes its dates of operation and a map showing its precise location. Users can retrieve the data either in tabular form for numerical analysis, or in easily read graphs (Figure 4.3). The data are the highest-quality information available about stream flow and often are used in engineering, scientific, planning, and legal studies.

The indicator measures of greatest importance for monitoring river hydrology to evaluate a possible dam removal include annual peak flow, annual low flow, annual mean flow, and real-time flow data. Annual peak flow is the highest discharge recorded in each year of record, and it provides a quantitative assessment of floods. The annual low flow is the

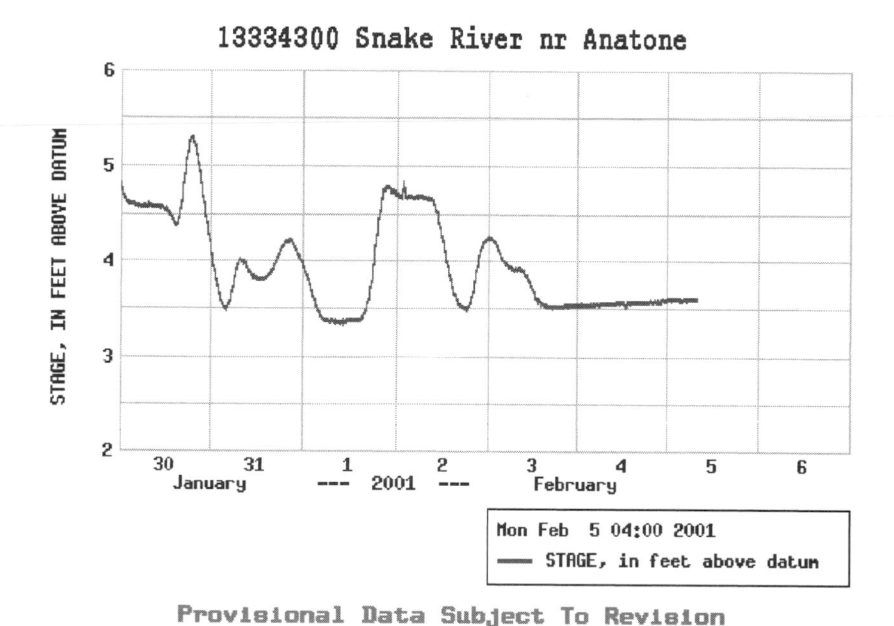

Figure 4.2 This graph shows an example of the changes in depth of flow that accompany rapid changes in discharge on the Snake River near Anatone, Idaho. See Figure 4.1 for a similar graph showing the changes in discharge. *Source*: U.S. Geological Survey (http://www.usgs.gov/water).

lowest discharge recorded in each year of record. The annual mean flow is the average of all the daily flows for each year; it provides a measure of water yield. An examination of the tabular data for peak flows provides the date on which the highest flow occurred in each year and addresses the timing issue. Many gage sites produce real-time information transmitted to the main USGS facilities in Reston, Virginia, and these data can be examined to determine short-term ramping rates (Figures 4.1 and 4.2).

SEDIMENT

Sediment Quantity

Rivers transport more than just water. Sediment transport and deposition occurs under entirely natural conditions. The construction and operation

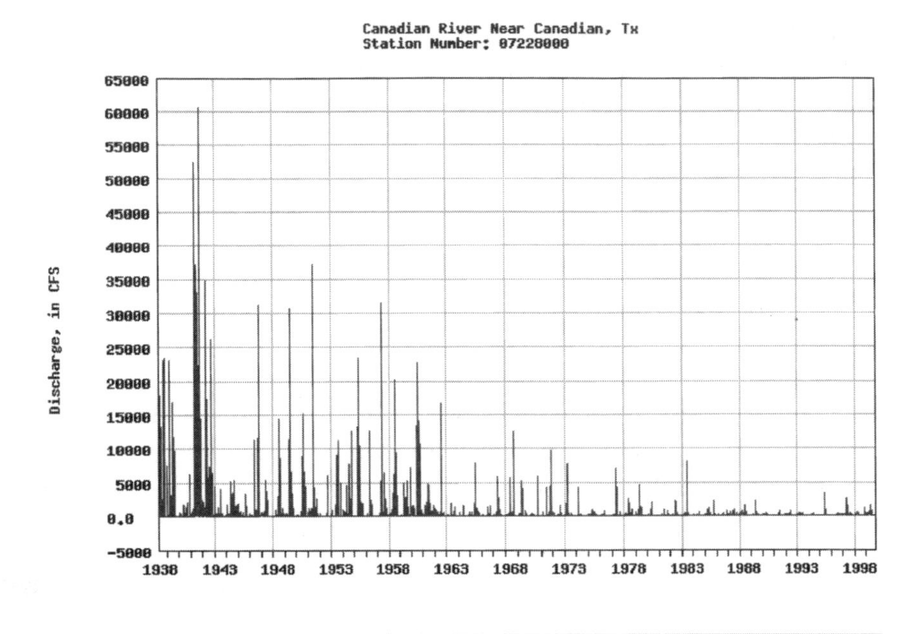

Figure 4.3 This graph is an example of historical (1938–1999) daily discharge data for the Canadian River near Canadian, Texas. Upstream storage reservoirs and diversions contribute to the decline in flows after the early 1960s. *Source*: U.S. Geological Survey (http://www.usgs.gov/water).

of dams affect the dynamics of sediments in rivers to a substantial degree, and dam removal is highly likely to involve issues related to sediment. Of the many physical outcomes related to dam removal, sediment erosion, transport, and deposition are likely to be among the most important. Dams trap sediment that enters their reservoirs because the turbulent downstream flow of water is halted temporarily. Sediments that are relatively large, such as gravel and cobbles, are deposited in deltas at the upstream ends of reservoirs, whereas smaller sediment particles, such as silt and clay, are carried farther into the pool area before they drop out. Most dams and reservoirs (excluding run-of-river dams) trap 95 percent or more of the sediment that enters them from upstream. Water released from dams is therefore relatively free of sediment, and downstream reaches do not receive the input of material that occurred before dam installation. As a result, the clear water erodes available sediment from the channel below the dam, winnowing away the finer material and leaving behind an "armor" of coarse particles too large to be entrained. Bank erosion in such cases is also common. The removal of dams restores the throughput of sediment and reconnects the various river reaches together in a continuous sediment transport system.

Sedimentation is considered by dam operators to be a problem in 25 percent of all reservoirs associated with hydroelectric projects, according to the results of a 1996 survey (Dixon, 2000). Sediment problems occur across the complete range of dam sizes and reservoir capacities when sediments occupy reservoir volume intended for water supply. As sediment fills reservoirs, it reduces storage capacity and the useful life of dams for hydropower generation. Sediment adversely affects dam operations by clogging power intakes, outlet works, and spillways, while also limiting recreational use of the reservoir by filling in surface areas. Because many dams create habitat for fish and wildlife, sedimentation also adversely affects these features. Sedimentation in the reservoir, initiated by the maintenance of a pool that triggers deposition, extends the adverse effects. Finally, the sediments themselves may accentuate chemical pollution, as outlined below. Matilija Dam in California is an example of a case in which history and the dam removal decision are closely connected with sediment (Box 4.1).

Sedimentation in a reservoir follows a general pattern.* Coarse sediments drop from the stream flow that enters the reservoir headwaters or backwaters area, creating a delta accumulation. If the reservoir contains water that is layered according to temperature, sediment-rich water containing relatively fine material (silt) may create turbidity currents issuing from the entry point of the stream into deeper reservoir waters. The resulting deposits consist of silt layers on the floor of the middle portion of the reservoir. The finest sediment (clay) is in suspension in the water and slowly settles out throughout the reservoir, including areas at the upstream face of the dam.

Rivers transport sediment from eroding landscapes to ocean and lake basins, temporarily storing significant quantities of sediment in floodplains, alluvial fans, and deltas along the way. Rivers that are in slowly changing, nearly equilibrium conditions exhibit a delicate balance among

---

* For more information, including researched examples, see Hakanson and Jansson (1983) and McManus and Duck (1993).

Box 4.1 Sediment Problems Associated with Matilija Dam in California

Matilija Dam exemplifies dams that are candidates for removal because of sediment problems. Since the dam was constructed in 1947 on Matilija Creek, a tributary of the Ventura River in Southern California, its reservoir has been filling slowly with sediment. Today, 6 to 7 million cubic yards of sediment lies trapped behind the structure, reducing the original storage capacity by over 90 percent. The reservoir is expected to be filled completely with sediment by 2010 (U.S. Bureau of Reclamation, 2000).

Because Matilija Dam traps the majority of coarse sediment normally transported during large floods, erosion problems have been severe16 miles downstream, along the famous surfing beaches of Ventura County. To preserve the beaches, protect coastal property, and maintain a coastal tourism industry that brought in an estimated $45 million to Ventura County in 1992 (State of California, 1997), costly measures such as beach nourishment, groins, revetments, and a seawall have been used. However, the beach structures are falling into disrepair, and multimillion-dollar projects are necessary to maintain them. Estimates show that up to 70 percent of the 50 years of sediment trapped behind Matilija Dam is suitable for placement on beaches, an amount sufficient to widen all south Ventura County beaches by 30 feet (Marx, 1996–97). Several studies examining the possible removal of Matilija Dam have been completed recently or are continuing.

Photo courtesy of Sarah Baish

To combat erosion, hard structures such as groins and seawalls have been constructed on Ventura County beaches, shown here in 2001.

discharge of water, discharge of sediment, and channel geometry (Leopold, 1994). Important human-induced changes to that balance include accelerated erosion in upland watersheds caused by agricultural practices, logging, recreational activities, and urban development; as well as changes to the river channel system, such as channelization and dam building. Dam removal is also likely to induce changes in the sediment transport and storage system.

Sediment is a pollutant because unwanted deposition fouls engineering works, and because artificially high turbidity (sediment suspended in flowing water) clouds otherwise clear water, degrading the quality of aquatic habitats for plants and aquatic animals, including fish. In 1998, sediment was the pollutant most often identified by states in their reporting of problems associated with the Clean Water Act (Federal Energy Regulatory Commission and Electric Power Research Institute, 1996). Sometimes, however, exceptionally clear water is also a problem, because native fishes in regions such as the interior Southwest have evolved in sediment-rich environments. Their populations decline in rivers dominated by the clear water released by dams (Minckley, 1991).

Dam removal involves potentially significant changes in the river's sediment system because the reservoir basin behind the dam is likely to contain quantities of sediments that would not be there if the dam had not been built. Dams form efficient sediment traps until their pool areas are completely filled with sediment. The amount of stored material depends on the size of the reservoir, rate of sediment supply from the upstream watershed, and length of time the structure has been in place. Many small, run-of-river dams and low diversion works have pool areas completely filled with sediment within a few years of their construction, whereas medium-sized and large dams typically have reservoirs only partly filled with sediments. In some cases, filling of the reservoir is so great that the water storage capacity is eliminated, as is the case for some dams slated for removal such as Rindge Dam (Box 3.1, p. 85) and Matilija Dam in Southern California (Box 4.1).

An important management question in decisions related to dam removal is the fate and quality of the sediments stored behind the dam. If the dam is removed, how much of the stored sediment will remain in place, and how much will be eroded by the flowthrough of water and be passed downstream? Are there any contaminants in the sediment that will pollute the river downstream if they are released? For most small, run-of-river dams, the majority of stored sediments are likely to be washed away by the river after the dam is removed, but for structures with large reservoirs designed for water storage, a problematical amount of sediment is likely to remain in place. Tim Randle and Gordon Grant, investigators attempting to predict the amount of sediment that will remain after the impending removal of dams on the Elwha River in Washington State, estimate that about half of the stored sediment is likely to remain in place. Mathematical models of the Snake River indicate that, if four major dams were removed there, 65 to 85 percent of the coarse sediment and about half of the fine sediment stored behind the dams would remain. In a potentially instructive case on the Gila River in Arizona, a breach of Gillespie Dam by a 1993 flood so far has resulted in the movement of large amounts of sediment stored behind the structure, but at least half the material remains in place. Although there is likely to be great variability from case to case, these and other instances show that much sediment is likely to remain in the old reservoir area after a dam is removed, and those sediments, whether contaminated or not, need to be taken into account in any successful management plan. They might be removed from the site or stabilized in place, but they cannot be ignored.

The existing sediment models were developed to predict the transport of extensive sediment deposits located in reservoir basins behind dams. Most sediment transport models are rather primitive from the standpoint of predicting downstream effects of reservoir evacuation; they do not handle the mixed grain sizes and staged export of materials very well. Models for sediment transport would need to be reviewed and modified to be useful in predicting potential sediment redistribution following a dam removal.

The sediments that are released by dam removal are carried downstream by the river flow, triggering a range of outcomes that may require management decisions. Fine sediments are likely to cause increased turbidity in waters downstream from the structure, and they eventually may be deposited as island, bar, or beach material along the stream below the dam site. If the river drains to coastal areas, these materials may be deposited in coastal wetland zones or transported by longshore currents to beach locations. Coarse sediments also may be eroded from the reservoir deposit, but they are likely to travel shorter distances in average flow conditions. They may be mobilized only during flood events, and when they are deposited, they may form bars of coarse material along the length of the channel or rapids across it. Deposits of fine and coarse materials downstream from sites of dam removals may create new ecological niches, which may be desirable from a river management standpoint or may supplant other, more desirable niches such as backwater areas and pools. The effective prediction of outcomes depends on reasonable estimations of the amount of material likely to be removed, understanding of the geomorphic and hydrologic behavior of the channel, and accurate hydraulic assessment of the post-removal river flows.

Because sediments released from the reservoir area of a dam that is removed migrate downstream, river managers may need to consider the consequences of these sediments eventually being stored behind the next dam downstream, if one exists. Most rivers have a series of dams on their channels, so the release of sediments in one part of the multi-dam system is likely to result in partial redistribution to other downstream dam sites. Assessments of likely outcomes of dam removal need to adopt a river-basin scale of analysis in estimating the potential issues associated with remobilized sediments.

In addition to the issue of sediment quantity, a further consideration in remobilized sediments is the effect on downstream particle sizes in the channel bed. Many salmonid fishes, for example, require a particular grain size for bed particles in their spawning areas, and if an upstream dam is removed, changes may occur in bed particle sizes. The covering of coarse bed particles by newly released fine material, for example, may decrease the usefulness of spawning areas. More research is needed on the effects of sediment releases of different volumes, grain sizes, and rates on downstream channels. This is the fundamental physical problem associated with dam removal and is linked to the biological response. On the other hand, the draining of reservoirs and evacuation of fine material may uncover previously drowned spawning beds by stripping away the fine sediment, with additional useful areas made accessible upstream by the elimination of the barrier to fish passage. Although it seems likely that fine materials will be flushed through the channel system downstream from most dam removal sites, the issue needs to be explored thoroughly during the decision-making process.

The quantity of sediment discharged past some gaging stations is available from the USGS as part of its stream gaging efforts, but the number of sites producing sediment data, about 1,600, is only a portion of the total gage system (Figure 4.4). The data, in a form similar to the water data, are available online (http://water.usgs.gov/owq.html).

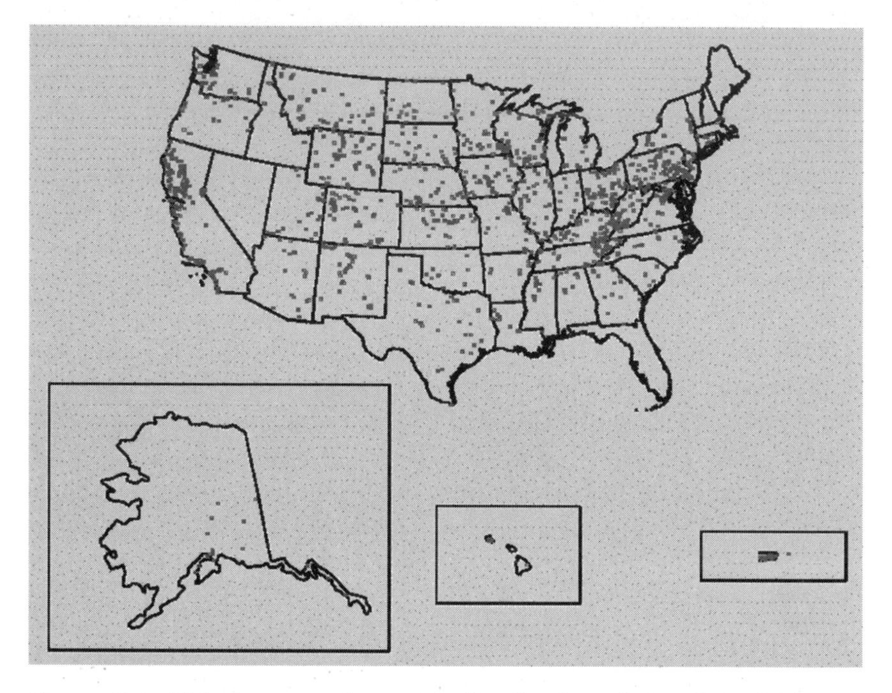

Figure 4.4 This dot map shows the distribution of sites for which sediment discharge data are collected for rivers. *Source*: U.S. Geological Survey (http://www.usgs.gov/water).

Sediment Quality

The chemical quality of the stored sediment is an important issue. Contaminants such as heavy metals, radionuclides, herbicides, and pesticides often are dissolved in river water, but they precipitate out of solution and are adsorbed onto the outer surfaces of sedimentary materials. The concentrations of contaminants in sediment are, therefore, often many times higher than they are in the water. These precipitation and adsorption processes are common in reservoirs, such that the remobilization of the sediments during dam removal presents downstream environmental risks. A proposal to remove low dams along the Blackstone River of Massachusetts was abandoned in the early 1990s because the sediments behind the dams were contaminated with heavy metals derived from manufacturing. The release of these sediments might have polluted downstream river and coastal habitats (Graf, 1996).

Concentrations of individual contaminants are useful indicators of sediment quality, and standards are available for assessing risk (Solomons and Förstner, 1984). A chemical assessment of sediments before removing a dam can define potential pollution problems, but a measurement of concentrations of individual pollutants requires laboratory analysis. Existing data on sediment chemical quality are available but scarce, and many river reaches and reservoirs have not been assessed on this basis. Sources of sediment quality data include state game and fish departments, along with state environmental quality departments that monitor particular locations either as short-term projects or part of general environmental protection programs.

Specific sediment quality standards are not commonly applied in the United States, and there are no general policies regarding acceptable limits for the concentrations of many pollutants in sediments. Some general guidelines for allowable limits are available for many chemical compounds, including herbicides and pesticides (Sittig, 1980), and these limits can be used in the assessment of sediments in reservoirs behind dams that might be candidates for removal. Heavy metals are common in sediments behind small, run-of-river structures because these dams tend to be placed in areas that were once manufacturing centers. Metals are also of concern behind tailings dams in mining areas. For these reasons, proposed classifications for concentrations of metals may be useful to decision makers in dam removal cases (Table 4.1). These classifications have not been adopted as formal policy but are supported by scientific investigations focusing on the biotoxicity of metals. The USGS is conducting a major sediment evaluation at Engelbright Dam, with full recovery, datable sediment cores.

GEOMORPHOLOGY

The hydrology and sediment systems of rivers build and shape the landforms that make up the river landscape. The geomorphology of rivers refers to these physical forms and processes that underlie the biological system. In some areas, vegetation plays a major, interactive role in landform development in and along rivers. Vegetation adds adhesive properties to riverine soils and enables them to resist erosion, and stems and leaf structures add hydraulic roughness to channel margins and floodplains. The installation of dams alters hydrology and sediment processes so that

Table 4.1 Proposed Classification of Sediment Pollution[a]

Element	Unpolluted (ppm)	Moderately Polluted (ppm)	Heavily Polluted (ppm)
Mercury	<1	Not defined	>1
Lead	<90	90–200	>200
Zinc	<90	90–200	>200
Iron	<17,000	17,000–25,000	>25,000
Chromium	<25	25–75	>75
Copper	<25	25–50	>50
Arsenic	<3	38	>8
Cadmium	Not defined	Not defined	>6
Nickel	<20	20–50	>50
Manganese	<300	300–500	>500
Barium	<20	20–60	>60

Source: Adapted from Baudo et al. (1990) based on research by Gambrell et al. (1983).
[a] This classification has not been adopted in formal policy, but is supported by scientific investigations focused on biotoxicity of the metals.

the downstream system responds with associated adjustments. Generally, the greater the storage capacity of a dam, the more extensive are its downstream geomorphic impacts. The most important of these impacts include channel shrinkage, deactivation of floodplains, changes in channel pattern, and loss of complexity.

Channel shrinkage is common downstream from dams because the structures often reduce annual peak flows. These peak flows are the "channel forming discharge" in many systems. Channel forming discharges are those flood flows that are efficient at moving sediment and shaping the channel, occurring about once per year or every two years. If the channel forming discharge is made smaller by flood control measures in dam design, the channel responds by also becoming smaller. Sediment accumulates on the bed of the channel or is deposited laterally along the channel side, resulting in reduced overall channel size. The ecological implications of these changes include the loss of aquatic habitat for fishes. In many cases, the total amount of available space in a channel is greater before the installation of a dam than afterwards, as exemplified by the Green River in Utah (Grams and Schmidt, in press).

In humid regions, a more appropriate measure is "bank-full discharge." Bank-full discharge is the level of discharge that is just sufficient to fill the channel to the tops of its banks, where additional water would spill out of the channel and onto the floodplains. This bank-full level is the most efficient transporter of sediment, occurs about once every year or two (at a time often referred to as the annual flood), and is a direct connection between the hydrology and geomorphology of rivers (Leopold, 1964). Williams (1978a) found that the bank-full measure was less reliable as an interpretive tool in dry land streams than in humid areas. Nonetheless, bank-full is a measure often used to describe rivers (Rosgen, 1994). The strength of the connections among annual flood, bank-full discharge, and channel size is variable from place to place (Miller and Ritter, 1996), but it is a useful guideline for explaining the effects of dams and anticipating the likely outcome of dam removals. The shrinkage of channels that occurs with the installation of dams is likely to be reversed if dams are removed, all other factors being equal.

The most useful indicator of the size of the river geomorphic system is channel width. Channel width, along with depth, gradient, hydraulic roughness, flow velocity, water discharge, sediment discharge, and sediment size, is a primary determinant of channel processes (Leopold, 1964; Leopold et al., 1994). Of all these variables, width is the most responsive to changes in the hydrologic behavior of rivers, as indicated by its relationship to discharge as defined by hydraulic geometry, a set of equations that relate physical and hydrologic properties of the river to each other (Leopold et al., 1964). Width has the additional advantage of being relatively easily measured in the field. For small streams, direct tape measurement is possible, and for larger streams, infrared ranging devices enable one person to make accurate bank-to-bank measurements with little technical support. All streams except the smallest ones have widths easily measured from aerial photographs.

Aerial photography, which is useful for assessing all the geomorphologic indicators described here, is widely available. If purchased from the federal government, the cost is often $10 or less for a single image that shows an area of several square miles, including up to 5 miles of river length. The most extensive source of aerial photography is the USGS EROS Data Center in Sioux Falls, South Dakota, which is accessible online (http://edcwww.cr.usgs.gov) (Figure 4.5 provides an example). The data center houses a vast array of federally obtained aerial photography, including historical images from the military, Coast and Geodetic Survey, and all the mapping photography made as the Geological Survey carries out its major mission of creating topographic maps for the nation. The

Figure 4.5 This example of aerial photography available from the U.S. Geological Survey's EROS Data Center is the type of image that provides geomorphologic information, including channel width, sinuosity, and pattern, as well as floodplain dimensions. The image is of terrain in southern Minnesota. *Source*: EROS Data Center (http://edcwww.cr.usgs.gov).

center contains at least one image for every area of the United States and has numerous dates of coverage for most areas. Table 4.2 reviews additional typical sources of aerial photography. Channel width measurements taken from photographs, although not as accurate as ground-based sur-

Table 4.2 Typical Sources of Aerial Photography for River Analysis and Monitoring

Type of Source	Institution
Federal Agencies	Geological Survey, EROS Data Center: small-scale aerial National Aeronautics and Space Administration satellite imagery from a variety of sources of the entire United States
	National Park Service: areas that include national park lands
	Forest Service: areas that include national forest land and nearby areas
	Bureau of Land Management: areas that include BLM land and nearby areas
	Fish and Wildlife Service: areas in and near wildlife refuges
	Natural Resource Conservation Service, formerly the Soil Conservation Service: repetitive coverage of agriculture, usually once every few years
	Bureau of Reclamation: larger streams and rivers in the western United States
	Army Corps of Engineers: local coverage of flood control project areas
	Department of Energy: areas on and near DOE facilities
	National Archives: historical images, including Soil Conservation Service photography dating to the 1930s
	Department of Defense: areas on and near military bases
Tribal Governments	Native American tribal governments for areas on and near Indian reservations
State Agencies	Land Department
	Environmental Quality Department or Department of Natural Resources
	Department of Transportation, especially highway divisions
	Game and Fish Department
	Water Resource Department
	Parks Department
	State Lands Department
	Industrial Development Commissions

(*continued*)

Table 4.2 *(Continued)*

Type of Source	Institution
County Agencies	Planning Department Highway Department Parks Department County Tax Assessor's Office
Town and City Agencies	Planning Department Water and Sewer Department Engineering Department Streets Department
Other Public Entities	University and college libraries University departments of engineering, geography, geology, life sciences, ecology State, county, city, or local historical libraries
Private Businesses	Electrical power companies Television cable companies Water companies Telephone companies Aerial photography companies

veys, are nonetheless precise enough to monitor changes resulting from dam installation and removal.

Channel width also is easily determined from engineering or scientific cross sections surveyed for planning, construction, or research purposes. These cross-sectional surveys are sometimes difficult to find, but they exist in surprising abundance. Government agencies involved with bridge construction, for example, almost always survey cross sections of the streams spanned by the bridges, and resource management agencies often survey stream cross sections taken for water, species, or land management purposes. Federal, state, and local highway departments are often useful sources for survey data. The Federal Emergency Management Agency (FEMA) contracts with civil engineering firms to conduct detailed surveys of many streams and rivers to determine the extent of the active channels and 100-year floodplains. The resulting cross sections and highly detailed topographic maps are available from FEMA online (http://www. fema.gov/maps).

The area of active floodplain in a given reach of channel is affected directly by the installation of dams, because most dams reduce the magnitude of flood discharges. As a result, the occasional floods are smaller than in pre-dam periods, and the area outside the channel that receives water, sediment, and nutrients in such events is less extensive. Thus, dams not only induce channel shrinkage, but also induce the shrinkage of active floodplains. Social and economic activities then encroach on the deactivated floodplain because people get the impression that they are safe from flooding. When rare, very large floods occur, they exceed the control capacity of some dams, resulting in the inundation of floodplain areas occupied by agriculture, industries, and residences. Additional protection for such properties in the form of levees further disrupts the once-active floodplain surface. The removal of dams is likely to result in higher occasional flood events and reactivation of floodplain surfaces (i.e., through periodic flood events large enough to flow over the floodplain, adding sediments and nutrients or eroding them away). Levees may restrict the out-of-channel flow of water temporarily, but sedimentation in the channel or breaches in the levees eventually result in inundation of the floodplain. For this reason, the reactivation of floodplains is incomplete, and additional measures dealing with the levees are sometimes required. For an example of the complex interactions among channels, floodplains, dams, and levees at a variety of scales, see Scientific Assessment and Strategy Team (1994). This assessment covered the midwestern United States, but the conclusions are broadly applicable.

In addition to changing feature sizes, the installation of dams simplifies rivers by reducing their geomorphic complexity. Although these effects are not yet well studied, the hydrologic and sediment adjustments caused by dams produces channels that have fewer complicated patchworks of different landforms than did the pre-dam arrangements. The loss of high ranges of flows and peak flows produces fewer islands, bars, beaches, and temporarily abandoned channels, so that there are fewer ecological niches. The removal of dams may reverse these changes, but there is no scientific research available to inform decisionmakers about this issue.

The frequency of islands, bars, beaches, and abandoned channels along a given length of channel provides a rough indicator of the complexity of the geomorphic system. If the numbers of such features are counted over time, the adjustments of the river can be traced to either installation or removal of control structures. The easiest way to determine the frequency of features along a channel is to assess aerial photographs, although for many systems, ground inspection or ground-based photography is also useful.

The installation of dams also affects downstream channels by changing their sinuosity and pattern. Sinuosity, in this case, refers to the degree of meandering of the low-flow channel. Because dams often reduce peak flows, and because the sinuosity of channels is controlled by peak discharges, channels respond to this hydrologic effect of dams. Generally, when flows are reduced and sediment loads decline, the channel becomes more sinuous, and the stream increases its winding characteristics. An additional adjustment observed in rivers in the Plains and western United States is a change in pattern from braided to single-thread geometry. A braided channel has many islands and bars with numerous sub-channels intertwined with each other (Figure 4.6). This geometry was common under unregulated conditions, with great ranges of discharge occurring over brief periods. In north Texas, for example, stream gage data show that unregulated rivers there have annual peak discharges that are 40 times the magnitude of the mean annual flow. Braided channels represent an accom-

Figure 4.6 A typical braided stream channel, Canyon Largo in the San Juan River Basin of northwestern New Mexico. *Source*: U.S. Geological Survey, http://water.usgs.gov, Aug.14, 2001.

modation to such radical variation. The imposition of dams, however, with the capability to smooth variations in discharge and eliminate very high flows, produces an entirely different hydrologic regime that often is conducive to maintenance of a single-thread channel. The removal of dams is likely to result in more braided conditions, especially if the bank materials have a low degree of cohesion (Schumm, 1977). A lack of woody debris and actively growing vegetation often causes braided conditions.

Sinuosity and braiding can be measured. The most common measure of the sinuosity of the low-flow channel is the ratio of the straight-line distance along the valley length between two points on the channel, to the along-channel distance between the two points (Leopold et al., 1964). A sinuosity ratio of 1.0 indicates a perfectly straight channel, whereas a ratio of 2.0 indicates a channel that is so sinuous that its length is twice as long as the straight-line distance between two reference points. Most river channels, including those in an unregulated condition as well as those subject to the influence of dams, have sinuosity ratios ranging between 1.1 and about 2.0. A useful measure of channel pattern is a braiding index proposed by Brice (1960). The braiding index is equal to the value of twice the total length of bars in a channel reach divided by the length of the reach itself. Investigators attempting to assess present conditions or historical changes in channels need to measure the landforms and calculate the sinuosity ratio and braiding index from aerial photography or maps, using the sources outlined above for the other geomorphic and hydrologic measures.

In summary, some of the most common, interrelated adjustments of the geomorphology of stream channels are controlled strongly by discharges of water, woody debris, and sediment. Knighton (1998) provides general descriptions of these associations, and Wohl (2000) offers insights into the processes and forms for mountain and canyon streams. Any changes to the discharges of water and sediment, such as those that occur with the installation or removal of dams, are likely to result in consequent adjustments in channel slope, particle size on the bed, channel depth, and channel width. The most likely general changes resulting from dam removal are increased water and sediment discharges resulting in decreased channel gradients, increased depths, and increased widths downstream. Initially, particle size may increase through the erosion of bed materials, but eventually bed materials may become finer as fine materials are released from behind the dam. These likely directions of change depend on a sequence of events that may be related to how a dam is removed, whether all at once, in stages, by progressive notching, or with staged drawdowns. The predications may vary from one case to another.

The state of geomorphic and sediment transport science for use by decision makers in dam removal cases, especially those involving small and medium-sized structures, is problematic. Extensive theory and model-based approaches are available for estimating the expected outcomes of dam removal. Hydraulic models and sediment transport models, for example, are widely available in the form of computer programs used by agencies and consulting firms (e.g., Simons and Sentürk 1992; Yalin 1992). Their application to situations involving dam removal has been limited, however, so this experience base needs to be weighed carefully. On the other hand, empirical research on the actual effects observed in dam removal cases is quite rare. Guidance on what to expect in terms of river channel change downstream from removed structures is, therefore, often lacking, and the decision maker is forced to rely on the judgment of a geomorphologist or hydrologist rather than on the more traditional scientific literature.

In 1997, as part of its efforts to improve the utility of federal environmental monitoring efforts, the White House Office of Science and Technology Policy asked The Heinz Center to identify a set of indicators for use in characterizing the state of the nation's ecosystems, using a nonpartisan, scientifically grounded process. The resulting report, to be issued in 2002, recommends improved collection of data on the extent of upstream effects of dams (i.e., inundation) and recognizes the need to quantify downstream effects as well. To help meet this need, the report recommends the development of quantitative tools and programs to monitor stream habitat quality. Such indicators and programs would measure changes in critical stream attributes resulting from a variety of causes, including the downstream effects of dams (The Heinz Center, in press).

WATER QUALITY

Dams have substantial effects on water quality because they alter the normal hydrologic behavior of rivers, which in turn changes the physical and chemical dynamics of the water. Among the most important potential changes resulting from the imposition of dams are oxygen depletion, temperature modification, changes in acidity, supersaturation of gases, elevated nutrient loading, increased salinity, and changes in contaminant concentrations in water and sediment. When nutrient-laden water enters a reservoir, some of the nutrients precipitate out of solution and become part of the sediment on the floor of the reservoir. This also happens with herbicides, pesticides, and heavy metals. Thus, the reservoir is a cleanser for water users, because the water released from the reservoir is lower in contaminants than the water entering it from above. However, these contaminants wind up in reservoir sediment. If a dam is removed, the sediments are remobilized and can carry their contaminant load downstream, causing a general decline in water and sediment quality (Petts, 1984).

Oxygen depletion occurs in reservoir waters because vegetation is inundated and decomposes in newly formed lakes, processes that use large amounts of dissolved oxygen from the water (McCully, 1996). Eventually, a new equilibrium is achieved, but usually the amount of dissolved oxygen in water released from reservoirs is less than that found in free-flowing rivers, which continually inject oxygen through turbulence. In reservoirs with substantial depth, stratification can create oxygen-poor conditions that may produce anaerobic water in the deeper areas, and if these waters are released to downstream areas, the water flowing below the dams is severely lacking in oxygen.

Dam installation may lead to decreases or increases in the water temperature of rivers downstream. The significance of this observation is that these changes may be reversed if dams are removed, with the expected changes resulting from either installation or removal depending on the characteristics of the reservoir and the withdrawal structures that take water from the lake. Reductions in release water temperature are common if the intake of the water is from lower levels of the reservoir, because cold water sinks to the bottom of lakes and is not circulated to warmer areas high in the water column. Strongly developed stratification preserves this cold water, and if it is released to downstream areas through bypass tubes, penstocks, and turbines, it substantially affects the downstream temperature regime (Petts, 1984). The effects depend on residence time of water in a reservoir. Because of their relatively small volume at any one time, unregulated rivers have large temperature ranges that respond to seasonal, synoptic weather, and diurnal changes (Walling and Webb, 1996). In warm climates, the river water is also warm most of the year. Releases from dams replace these changeable conditions with cold water characterized by only minor fluctuations in temperature, conditions very unlike the original pre-dam conditions. Some dams release warm water in winter. Rivers with dams impounding reservoirs experience fish habitat changes connected with temperature adjustments that are often unfavorable to native fishes accustomed to warmer waters (Stanford and Ward, 1991), but favorable to non-native or introduced gamefish (e.g., trout).

Temperatures increase in release water either when the reservoir is shallow or when the withdrawal structure is close to the surface of the lake. In shallow reservoirs, seasonal warming heats the slow-moving waters to temperatures that are higher than those experienced in free-flowing streams, a situation very different from the stratification and isolation of cold, deep water in deep reservoirs. Withdrawal structures that are situated near the surface of a reservoir behind a dam also may supply warm water to downstream areas, because the warmest water is usually found near the lake surface. In any event, these temperature changes represent adjustments from previously unregulated flows, and the removal of a dam is likely to bring about readjustments in temperature for downstream reaches.

Changes in acidity occur in reservoir waters because of evaporation from the surfaces of artificial lakes. The waters entering reservoirs contain a certain amount of dissolved solids, but evaporation removes some of the water, leaving behind increased concentrations of dissolved solids, which in turn increase the alkalinity, or pH, of the remaining water. The most common dissolved solid in these cases is salt. The reservoir waters released through dams to downstream areas bring the increased salinity to aquatic plants and animals as well as riparian vegetation tuned to pre-dam low salinity conditions. The maintenance of native species is, therefore, more difficult with dams and storage reservoirs in place. River flow characteristics have a strong influence on dissolved solid concentrations in any case (Webb and Walling, 1996), and the alterations of river hydrology brought about by dams causes downstream changes in pH even if there are no increases in salinity in their reservoirs.

The supersaturation of reservoir waters with atmospheric gases such as nitrogen occurs because water is "buried" in reservoirs, where the increased hydrostatic pressures at the bottom of the reservoir force these gases into solution. If dam operations draw water from these lower levels for release, the supersaturated waters enter downstream reaches and strongly affect fish (Baumann et al., 1986). The release of water through turbines and penstocks contributes to this supersaturation (Petts, 1984), a circumstance that is hazardous for fish because they absorb the gases into their blood during respiration. As the gases come out of solution in a fish's bloodstream, the fish experiences a condition similar to the "bends" experienced by divers who surface too rapidly. The supersaturated gases cause the blood to bubble, a debilitating and sometimes deadly condition (National Research Council, 1996).

Dam installation may improve water quality downstream from a dam site, and removal of the structure may reduce water and sediment quality downstream. This process is the result of a complicated series of physical and chemical processes that affect nutrients, herbicides, pesticides, and heavy metals. When these contaminants enter a reservoir area dissolved in water, they often precipitate out and become associated with the sediment on the floor of the reservoir. As a result, water released from the reservoir may be of higher quality than that entering from above. However, if the dam is removed, the contaminant-enriched sediments are released and remobilized, potentially creating pollution problems downstream.

A reservoir is a sink for nutrients when a dam is in place and a source for nutrients when the dam is removed. Nutrient loading occurs in reservoirs in agricultural and urban areas because of runoff contributions. The application of fertilizers to agricultural lands at a large scale and to suburban lawns at smaller scales produces an abundance of nutrients in many watersheds that does not occur under natural conditions (Baumann et al., 1986). Runoff from these fertilized areas contributes large amounts of nitrogen, phosphorus, potassium, and other chemicals to the water flowing into reservoirs. The evaporation of reservoir water in dry land settings concentrates contaminants, especially metals and salts, in the remaining waters (Salomons and Forstner, 1984). These elevated concentrations of metals enhance the transfer of contaminants to sediments in the reservoir, so that water released from the dam has reduced concentrations of these materials. Salt, on the other hand, may remain in solution, and the increased salinity is passed through the dam to downstream areas.

Contaminant concentration in sediments increases in reservoirs because of the exchanges between water and sediment (Baudo et al., 1990; Horowitz, 1991). The concentrations of contaminants, especially heavy metals, are often two or three orders of magnitude greater in sediment than in the overlying water (Glover, 1964; Salomons and Forstner, 1984). Herbicides, pesticides, heavy metals, and radionuclides are transported into the reservoir system by flowing water, but once in the lake they begin to be adsorbed onto the surfaces of sediments suspended in the water and resting on the bottom. Thus, as stated earlier, the presence of a dam and its reservoir may serve to protect downstream areas from contaminated sediments, but the removal of the dam may serve to remobilize stored contaminated material. Such material needs to be removed and disposed of before dam removal if contaminants are at issue.

The effects of dams and reservoirs on dissolved oxygen, temperature, pH, supersaturation of gases, and nutrient loading are all reversed with the removal of the dams. However, in the case of nutrient loading, dam removal does not return river reaches to entirely natural conditions. An important outcome of dam removal is the release of sediments that contain contaminants in exceptional concentrations because of adsorption onto sedimentary particles. Even the sediments stored behind small, run-of-river structures need to be examined to determine their quality. For example, the removal of Waterworks Dam on the Baraboo River in Wisconsin entailed the physical removal and disposal of its stored sediments to avoid downstream dispersion of contaminants.

The indicators for water quality are direct measurements of temperature and pH, along with laboratory analysis of water samples to assess concentrations of chemicals such as organic chlorines (indicators of herbicides and pesticides), heavy metals, and radionuclides. Laboratories using chemical methods stipulated by the U.S. Environmental Protection Agency and Environment Canada best conduct these analyses. The use of approved methods ensures compatibility with previously collected and published measurements. A readily accessible database for water quality in rivers of the United States is the National Water Quality Assessment Program (U.S. Geological Survey, 2000). Near real-time and historical data for many of the nation's rivers are available online (http://water.usgs.gov/owq/data.html). The data can be downloaded in the form of tables for much, but not all, of the country (Figure 4.7).

CONCLUSIONS AND RECOMMENDATIONS

Removing dams can restore some of the most important aspects of physical integrity to rivers downstream. In addition to the effects of their reservoirs, which inundate terrain and ecosystems, dams affect physical integrity by fragmenting the lengths of downstream rivers, changing their hydrologic characteristics (particularly peak flows), and altering their sediment regimes by trapping most of the sediment entering their reservoirs. These effects translate into major changes in the downstream geomorphology of the river landscape, most critically through channel shrinkage and deactivation of floodplains. Water quality changes also alter the eco-

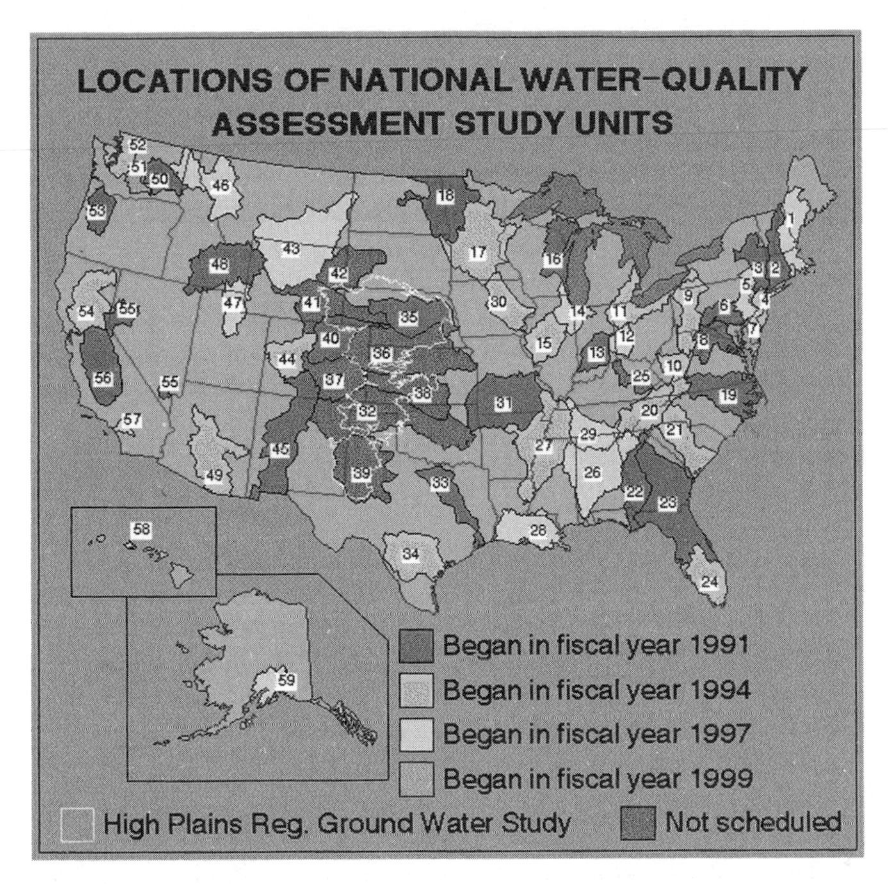

Figure 4.7 This map shows the distribution of basins for which highly detailed water quality information is available through the National Water Quality Assessment Program. *Source*: U.S. Geological Survey (http://www.usgs.gov/water).

system downstream. The removal of dams has the effect of reversing most of the undesirable changes, but it is unlikely to restore completely natural conditions because of other dams on the river and the multitude of other human-induced effects on streams, such as channel control and land use in watersheds upstream. The most important positive outcome of dam removal is the reconnection of river reaches so that they can operate as an integrated system, which is the basis of a river with physical integrity. Productive, useful ecosystems can result from dam removal, but outcomes may include many interrelated changes in the physical and biological components of the river. Dam removal results in the remobilization of sediments once stored in reservoirs, and some of these sediments may be high in nutrients or contaminated by pollutants. Therefore, planners and researchers need to undertake wide-ranging assessments of likely outcomes of dam removals that account for anticipated changes in water, sediment, landforms, vegetation communities, and fish and wildlife.

- **Conclusion:** Sediment processes are the most fundamental aspects of dam removal issues that are poorly understood. Water quality is important because of its human health and environmental dimensions; it is governed by extensive policies, yet outcomes of dam removal on water quality are poorly understood. Empirical data are lacking on river channel change downstream from removed structures.
- **Recommendation:** The panel recommends that the scientific community of river researchers provide (1) improved understanding of sediment quality and dynamics to provide a scientific basis for evaluating contaminated sediments, (2) improved understanding of the roles that dams and their potential removal play in water quality models, (3) empirically derived explanations of river channel change upstream and downstream from removed dams; and (4) a knowledge base of the likely fate of sediments and their contaminants downstream from removed dams.

- **Conclusion:** There is a glaring need in the science and decision-making communities for a geospatial database that provides accurate, readily accessible data about the segmentation of the nation's rivers by dams.
- **Recommendation:** The panel recommends that U.S. Environmental Protection Agency and/or U.S. Geological Survey should consider augmenting the existing national stream reach geographical data to include the location of dams and to allow better analysis and understanding of the segmented nature of the nation's streams and rivers.

- **Conclusion**: The quantity of sediment discharged is available from the U.S. Geological Survey as part of its stream gaging efforts. However, the number of gages producing sediment data is only a portion (1,600) of the total national gage system (6,600).
- **Recommendation**: The panel recommends that the U.S. Geological Survey maintain and extend its network of sediment measurement statistics throughout the total national stream gauging system.

5

Biological Outcomes of Dam Removal

Aquatic ecosystems include components ranging from an entire watershed to one-celled bacteria that are responsible for primary decomposition. Rivers are dynamic entities that undergo change and evolution, continuously creating, evolving, and realigning new aquatic habitats. An aquatic ecosystem is a complex continuum of habitats that include production zones, spawning areas, refugiums for various life stages of fish and metapopulations (Figure 5.1), migration corridors, feeding stations, and a plethora of unique microhabitats. The physical and biological processes of the river system define each ecosystem component.

This chapter explores the biological aspects of rivers that are relevant to decisions about dam removal. As is evident with respect to the physical and hydrologic aspects of rivers, scientists know a great deal more about the biological changes effected by the installation of dams than about those induced by dam removal. The chapter begins by providing a framework of levels of change and response and then discusses the fundamental contexts for restoration: spatial, temporal, and ecosystem contexts. Finally, this chapter reviews the various factors that affect restoration.

Aquatic ecosystems are the products of the dynamic relationship between the watershed and the biological resources that live in the river system. A river is the sum of its parts and often has been referred to as a continuum of ecosystems and processes (Figure 5.2) (Vannote et al., 1980). Rivers and reservoirs are shaped by inputs from the upstream watershed. Rivers and reservoirs exhibit different trophic relationships due to modified hydraulic dynamics and variables. A trophic hierarchy exists in rivers, building from the primary sources of energy, the algae and macrophytes; to the primary consumers; to, ultimately, the fishes and the

Figure 5.1 The species of fish that live in rivers include brook trout (top left); pike minnow (top right, an endangered species); Gila trout (bottom left, also endangered); and Virgin River chub (bottom right, endangered). Courtesy of the U.S. Fish and Wildlife Service, photographs by Duane Raver Art.

Figure 5.2 The aquatic environments on the free-flowing South Fork of the Cumberland River in Tennessee range from rapids to tranquil pools. A dammed river, in comparison, has less varied habitats. Photo courtesy of the U.S. Army Corps of Engineers.

issues associated with their distribution and abundance, life-history adaptation, and management. In river systems without dams, changes are often subtle unless a large, river-reshaping event occurs, such as a flood or massive land change (Poff et al., 1997; Richter et al., 1996).

The shape and size of a river is a function of the flow, quantity and character of the sediment in transport, and character and composition of the materials that make up the bed and banks of the river (Leopold, 1994). By affecting quantity and timing of water flow, flow velocities, water chemistry and biogeochemical cycling, dams change the dynamic relationship between the watershed and river and, consequently, affect the species that depend on the river and riparian area for their survival (Ligon, Dietrich, and Trush, 1995; Power et al., 1996). According to a recent report by the Wisconsin Department of Natural Resources, dams are among the most significant obstacles to restoring the biodiversity and integrity of riverine systems (Born et al., 1998).

A watershed perspective is especially helpful in visualizing and understanding the physical systems of rivers, as outlined in Chapter 4. However, watersheds also can be viewed as ecosystems, functioning collections of organisms and their inorganic support systems (Meehan, 1991). Ecosystem boundaries are difficult to map and interpret, but watershed boundaries are usually clearly defined, and they make useful margins for biological analyses involving water-related resources. Scientists and policymakers involved in dam removal decisions will find the recently published *Freshwater Ecoregions of North America: A Conservation Assessment* (Abell et al., 2000) especially helpful in this regard, because it focuses largely on biological resources and is compartmentalized according to watersheds and river basins. The maps and diagrams in the report are helpful in placing dam decisions in both a physical and a biological context.

POTENTIAL IMPACTS OF DAM REMOVAL ON AQUATIC ECOSYSTEMS

The placement of a dam and a reservoir on a river modifies the biogeochemical cycles both in the reservoir and downstream (Stanford and Ward, 1979). Dams immediately fragment the river system, leading to modified flows (i.e., water quantity, timing, and quality) and, subsequently, changes in the movement patterns, process times, available habitats for fish and macroinvertebrates, and ultimately has resulted in losses of biodiversity. The modification of a natural flow regime has direct, indirect, cumulative, and specific watershed-level impacts on an aquatic ecosystem (Poff et al., 1997)

Changes in the river's aquatic ecosystem responses are defined and controlled by the limnological events occurring in the upstream reservoir and further modified by the discharge regime from the dam (American Fisheries Society, 1985; EPA, 1989; Tyus, 1999; U.S. Department of Energy, 1994). To predict the effects of dam removal, it is necessary to understand how dams have influenced the downstream and upstream environments. The key aquatic ecosystem characteristics that can be used to assess the influence of dams as the downstream environment include the following:

- Modified substrates associated with the armoring of the streambed downstream and the subsequent reduction in the amount and availability of spawning habitats
- Loss of small-grained sediments necessary for transferring nutrients and for providing substrate for riparian and aquatic plants (Wilson, Gendhe, and Marston, 1988; Gresch, Lichatowich, and Schoonmaker, 2000)
- Loss of the ability to support nutrient and energy flow (Larkin and Slaney, 1997; Cederholm et al., 1999)
- Modified thermal regimes in terms of timing, ecological cues for life cycles, and total number of degree days necessary for development (Collier, Webb, and Schmidt, 1996; Vinson, 2001)
- Modified downstream aquatic saprophyte assemblage due to changes in sediment delivery, thermal conditions, seasonal floods, etc. (Voelz and Ward, 1991; Stevens, Shannon, and Blinn, 1997; Andrews, 1986)
- Modified macroinvertebrate species diversity associated with changed thermal cues, habitats, and timing of life history strategies (Vinson, 2001; Lehmkuhl, 1972)
- Modified fish assemblages associated with changes in habitats, introduction of non-native species, changing food bases, and modification of thermal and other water quality and quantity cues necessary to initiate specific life history cycles; these effects cause changes in the reproductive cycles and growth of young fish and result in the loss of effective migration ability among adult and juvenile fish (National Research Council, 1992, 1996; Petts, 1980)

Just as constructing a dam alters the natural environment of a river, removing a dam also alters the aquatic ecosystem below and above the structure. The changes are variable on both temporal and spatial scales. The ecological changes depend on the size of the structure and amount of water that it impounds; quantity and quality of sediment trapped in the reservoir; season and timing of the draining of the reservoir; native and non-native fish and invertebrate species that inhabit the reservoir; limnological conditions in the reservoir; and stability of the downstream river channel (Burns, 1991; Dynesius, Nilsson, 1994).

Dams create reservoirs, which are artificial bodies of water that create modified hydrologic, physical, and biological environments that differ from those provided by rivers. Depending on the size, watershed, and management of the reservoir, wetland habitats may be temporarily formed at the low area of the reservoir and the number of species may temporarily increase. Problems with reservoirs arise, however, as water levels fluctuate, resulting in direct impacts to wetland habitats (Bolke and Waddell, 1975; Heiler et al., 1995).

The removal of a dam has both short-term and long-term effects on a river's aquatic ecosystem and biodiversity. Biodiversity, short for biological diversity, is defined by the National Research Council (1997) as the variety of life found on the planet. Although dams can have some positive ecological effects (e.g., creating additional wetland habitat), removing a dam may increase the abundance and diversity of aquatic insects, fish, and other organisms (Doyle et al, 2000; Ward and Stanford, 1995; Malmquist and Englund, 1996; Doeg and Koehn, 1994; Camargo and Voelz, 1998). Wetlands surrounding the reservoir may be lost, but wetlands and riparian areas along the banks of the rivers may be restored. In addition, although water quality often is degraded immediately following a removal, the restoration of a river's natural flow may eventually result in improved aquatic habitat. Once a dam is removed and crucial upstream habitat becomes accessible, migratory fish populations (including endangered or threatened species) often rebound (American Rivers et al., 1999). The removal of a dam has a lasting impact, however, on some game species of fish (Shuman, 1995) (Figure 5.3). Dams usually change rivers from a state of constant flow to a more lake-like condition with standing water.

Different species of fish use specific habitats. Panfish, catfish, and largemouth bass are typical of the fish assemblages that are supported by reservoirs. Native fish species have evolved specific life history characteristics that allow them to survive and flourish in a flowing water environ-

Figure 5.3 Game species, such as these striped bass, may be affected by the removal of a dam. Courtesy of the U.S. Army Corps of Engineers.

ment, such as a river. With the removal of a dam, the fish assemblages supported by the reservoir are forced to change (Jennings, Forem, and Karr, 1995). Studies conducted on the fish assemblages on the Baraboo River in Wisconsin showed that the communities changed rapidly as the river reclaimed its natural flow dynamics (Catalano et al., in press). Within eighteen months after the removal, the number of fish species upriver from the former dam site increased from 11 to 24, according to a Wisconsin Department of Natural Resources survey. The number of smallmouth bass species, which cannot tolerate poor water quality, increased from 3 to 87 (American Rivers et al., 1999; Kennebec Coalition, 1999).

The response of an aquatic ecosystem following a dam removal may result in a different aquatic community than existed before dam construction (Wik, 1995; Travnicheck et al., 1995; Shuman, 1995). The pre-dam community may reappear only through active restoration activities, such as non-native fish eradication, habitat and substrate restoration, and watershed management. (In some cases, the pre-disturbance species may have been eliminated and upstream watershed processes modified.) The bottom line is that the pre-dam aquatic community likely has changed, through development and successional processes, in response to the natural and modified physiochemical environment, watershed, and habitat changes. These changes seldom are related directly to the dam in question, so dam removal by itself is unlikely to restore the exact ecological conditions that existed before human occupation of the floodplain (Box 5.1). It is possible to reach limited restoration goals with dam removal, especially in the reestablishment of fish passages (Box 5.2).

Box 5.1 The Unanticipated Impacts of Removing Fort Edward Dam in New York

The experience of removing Fort Edward Dam shows how complicated true river restoration can be. The project also demonstrates the need for comprehensive pre-removal environmental assessment studies. Constructed in 1898, Fort Edward Dam was a 586-foot-long, 31-foot-high hydroelectric dam on the Hudson River in New York. Its owner, the Niagara Mohawk Power Corporation, removed the dam in 1973 after a study concluded that it was a public safety hazard.

Although several studies and analyses were conducted before the removal, they were inadequate with respect to determining the full impact on surrounding areas, aquatic ecosystems, and navigation. Soon after the dam was removed, unanticipated water quality and navigational problems appeared, some of which continue to this day. For instance, the quality of the sediment trapped behind the dam was not analyzed sufficiently to discover the presence of polychlorinated biphenyls (PCBs) that had been accumulating from an upstream chemical manufacturing plant. The sudden release of these contaminants was catastrophic for the river's ecosystem, causing New York State to close the Hudson River to fishing in 1976 and the U.S. Environmental Protection Agency to declare a portion of the river a federal Superfund site in 1983. In addition, the sediment moved downstream and effectively blocked a large portion of the Hudson River navigation channel, a marina, several industrial sites, and other downstream areas. The channel's reduced capacity and restricted water flow also increased the flood hazard for the town of Fort Edward and created a public health hazard when untreated raw sewage released into the river began to stagnate.

These unanticipated impacts resulted in several lawsuits and millions of dollars in lost revenue (mainly for fisheries and navigation) in addition to the clean-up and restoration costs. Lessons from the Fort Edward dam removal have been incorporated into more recent dam removal decisions. The lessons include the importance of testing and analyzing both the quantity *and* quality of the accumulated sediment and determining the potential impacts of the sediment release and decreased water flow on the entire upstream and downstream environment.

Box 5.2 The Restoration of Butte Creek in California

Dam removal can be an effective part of a more comprehensive river restoration program. The 1998 removal of four water diversion dams and 12 unscreened water diversions on Butte Creek in Sacramento Valley, California, was the result of a decision based on both agricultural and ecological values. Rice farmers bordering Butte Creek traditionally eliminated rice stalks from the previous year's crop in their fields by burning. In 1991, air pollution from this activity became so problematic that an alternative had to be found, and the rice farmers turned to the river. By flooding their fields, they could accelerate the decomposition of the rice stalks. However, the California Department of Fish and Game was concerned that the unscreened water diversions might further harm Butte Creek's already threatened salmon population.

In 1987, only 14 spring-run chinook had been found spawning upriver of the dams in Butte Creek. Once California's most abundant salmon species, the spring-run chinook was listed as threatened under California's Endangered Species Act and was being considered for listing under the federal Endangered Species Act. If the chinook were listed, not only would the rice growers not be able to flood their fields, but commercial fishermen would not be able to fish on Butte Creek, and pumps on the creek, which supply Southern California with water, could be shut down. The California Department of Fish and Game, Western Canal Water District (which owned two of the dams), U.S. Department of the Interior, and California Urban Water Agencies worked collaboratively to remove the four dams and 12 diversions. The project's total cost was $9.13 million. Removing the dams and diversions restored 25 miles of Butte Creek to a free-flowing condition, and more than 20,000 adult chinooks spawned in Butte Creek in 1998.

Source: American Rivers et al., 1999.

AQUATIC ECOSYSTEM RESTORATION PLANNING

All restorations are exercises in approximation and in the reconstruction of naturalistic rather than natural assemblages of plants and animals with their physical environments (Berger, 1990). In some cases, rehabilitation may be a more descriptive term, because the management goals may be to repair damage to river processes and forms resulting from a dam or its operation. The removal of a dam provides an opportunity for a river to partly reconnect its watershed. Predicting what will happen to the aquatic ecosystem when a dam is removed is more complex than simply taking down a dam and letting "nature" take its course. Whether a dam removal project is judged a success or not will depend on the goals and objectives of the removal. As discussed in Chapter 3, these goals and objectives of dam removal need to be articulated as the first step in the decision-making process regarding dam removal. Legal considerations may force a decision to remove a dam for safety reasons, while the restoration of habitat for legally protected species may be the primary factor. The removal of Savage Rapids Dam in Oregon is one of many such cases (Box 5.3).

"Restoration" of rivers is a commonly stated goal in dam removal decisions, so decision makers need to be clear on the implications of using the term (National Research Council, 1992). Restoration is defined by the National Research Council (1992) as

> The return of an ecosystem to a close approximation of its condition before disturbance. In restoration, ecological damage to the resource is repaired. Both the structure and the functions of the ecosystem are recreated. Merely recreating the form without the functions, or the functions in an artificial configuration bearing little resemblance to a natural resource, does not constitute restoration. The goal is to emulate a natural, functioning, self-regulating system that is integrated with the ecological landscape in which it occurs. Often, natural resources restoration requires one or more of the following processes: reconstruction of antecedent physical hydrologic and morphologic conditions; chemical cleanup or adjustment of the environment; and biological manipulation, including revegetation and the reintroduction of absent or currently nonviable native species.

The National Research Council (1992) also notes that no restoration can ever be perfect. It is impossible to replicate the exact biogeochemical and climatological sequence of events over geologic time that led to the creation and placement of even one particle of soil, much less to restore an entire ecosystem. In developing restoration strategies, the recovery of an ecosystem to an approximation of its natural predisturbance condition needs to be pursued as the first goal. In many situations, this ideal may not be practical, physical, and biological as illustrated in Figure 5.4.

The shaded area represents an "envelope" in which the morphology and function of the ecosystem are considered acceptable and achievable under existing social, political, economic, and engineering constraints (NRC, 1992). The goal in this restoration scenario would be to transform

Box 5.3 Legal Considerations in the Removal of Savage Rapids Dam in Oregon

The decision about whether or not to remove Savage Rapids Dam, located on the Rogue River in Oregon, is an example of a process driven by concern for an endangered species and habitat restoration. This concrete diversion dam, which stands 39 feet high and 460 feet wide, was constructed in 1921 to divert water to farmers. The dam's owner, the Grants Pass Irrigation District (GPID), has known for years that the dam impedes the upstream and downstream passage of salmon and steelhead trout. Moreover, the dam no longer provides any flood control, storage, or power generation benefits. Twice, in 1994 and 1997, the GPID Board of Directors voted in support of dam removal, thus agreeing with the National Marine Fisheries Service (NMFS) to improve passage for the endangered fishes. In 1995, the U.S. Bureau of Reclamation completed a study that estimated the cost of refitting the dam to be less lethal to salmon could run as high as $21 million. In contrast, removing the dam and meeting local

The Savage Rapids Dam in 1999.

Courtesy of Waterwatch

the ecosystem, by the time the project is complete, from its present state to some point within the achievable envelope.

Spatial and Temporal Contexts

From a spatial perspective, rivers operate within a specific arrangement of the earth's surface that is delimited by watersheds (National Research Council, 1999). Watersheds are areas of land surface that contribute

Box 5.3 continued

water needs with modern pumps was estimated to cost only $13 million (Waterwatch, 2001).

The GPID did not remove the dam when new board members were elected. In response, the NMFS brought suit against the irrigation district under the federal Endangered Species Act (ESA) in 1998, because one of the salmon species affected by the dam is the ESA-listed coho. Several conservation, sportfishing, and commercial fishing organizations have joined the federal government in the pending lawsuit. Despite efforts by the GPID to save it, Savage Rapids Dam may be removed because of legal actions under Oregon water law and the ESA. Dam removal and replacement with pumps is the only permanent solution to the problem and the only solution that eliminates the GPID's ongoing liability for fish losses at the dam. It is also the only solution that guarantees that the GPID will receive an incidental take permit under the ESA and allow settlement of the ongoing litigation with the federal government. This permit is needed to ensure the GPID's continued right to operate its diversion system.

In October 2001, the governor of Oregon signed a consent decree that dissolved the state and federal lawsuits against the GDIP over the harm the dam has caused endangered coho salmon. The agreement calls for a new pumping system that will divert water into irrigation canals without disturbing the fish in the river to be installed by 2005, followed by dam removal by 2006 (Olson, 2001). However, the agreement depends on the U.S. Congress approving at least some funding for the project, estimated at $22.2 million. U.S. Senator Ron Wyden (D-Oregon) introduced the Savage Rapids Dam Act of 2000 (S. 3227) in the 106th Congress. However, there was insufficient time to pass the bill in the last session and no comparable bill has been reintroduced. Initial removal study funding of $500,000 has been appropriated (Grants Pass Irrigation District, 2001).

runoff—including water, sediment, and chemicals—to confined channels (Williams, Wood, and Dombeck, 1997). Small basins are nested within larger ones in a topographically defined arrangement, culminating in the watershed that constitutes the areas and supports river basins. The aquatic resource is organized, supported, and defined according to watersheds. The dams and their effects are best understood in a watershed context (Dynesius and Nilsson, 1994; Stanford and Hauer, 1991; Ward and Stanford, 1995). The removal of a dam strongly influences its immediate site and reservoir area and is likely to have effects far downstream. Removals

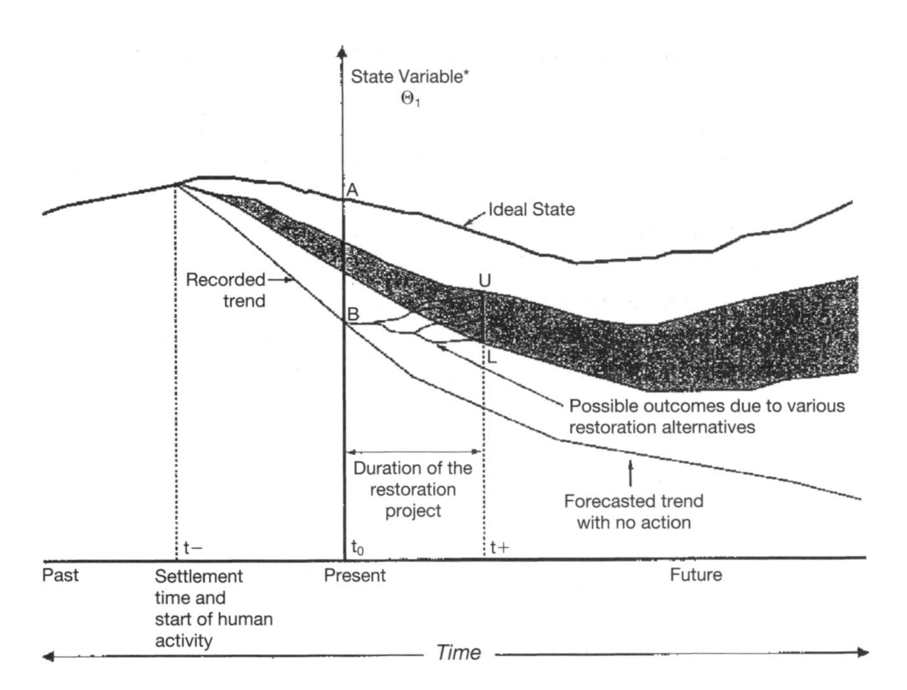

Figure 5.4 Schematic representation of a restoration scenario. *Source:* Reprinted with permission from National Research Council (1992).

of dams also may propagate effects upstream of the impounded reach by reconnecting headwater areas to aquatic organisms that can migrate upstream without an impeding structure and reservoir in place. Reconnecting the river's ecosystem will allow for retrieval and energy exchange (Hall, 1972; Wood and Armitage, 1997; Camargo et al., 1998; Hughes and Noss, 1992), sedimental redistribution (Petts, 1980; EPA, 1989; Tyus, 1999), and fish passage (AFS, 1985; Raymond, 1988; Burns, 1991;

Bates, 1993). Because dam decisions affect watershed-scale processes, the decisions often should be made within the same watershed context. The Conestoga River dams of Pennsylvania exemplify this approach (Box 5.4).

Restoration can involve passive or active processes or both. Passive restoration uses the natural river processes following their own timetable. Active restoration involves direct actions and management to assist in the restoration effort.

Aquatic Ecosystem Indicators of Restoration

Determining what an aquatic ecosystem restoration will look like is an essential first step in developing and implementing a credible dam removal program. The determination of ecosystem responses to restoration actions is complex and a criterion of immense importance in the dam removal process.

Evaluating of recovery patterns of aquatic ecosystems requires the selection and use of indicators that are characteristic of the specific aquatic ecosystem and that include the appropriate spatio-temporal scale of observation (Kelly and Harwell, 1990). Biotic and abiotic indicators commonly are used to evaluate aquatic ecosystem responses (Ward and Stanford, 1979; Shuman, 1995; Karr, 1981; Auble, Friedman, and Scott, 1994). Figure 5.5 is a graphic representation of possible functional end points.

An aquatic ecosystem's biotic response following dam removal needs to be evaluated at both structural and functional community levels. Most common macroinvertebrates and fish species are used as indicators of aquatic community responses. Structural criteria include the composition of the community assemblages in terms of attributes such as density, number of species, and species diversity, along with indicator and keystone species (Milner, 1994). Criteria typically include a comparison to pre-disturbance times or a reference community. Functional criteria refer to the response of the community as indicated by production, trophic and species equilibrium, and the existence of keystone species. A common methodology used to evaluate community function is the concept of biological integrity, which is described well by Karr (1994).

A biotic response includes both habitat and water quality. Physical habitat quality may include the amount of gravels for spawning, heter-

Box 5.4 Dam Removals in the Conestoga River Watershed in Pennsylvania

Decisions about dam removal on the Conestoga River demonstrate the value of a watershed perspective in the decision-making process. The Conestoga River and its tributaries in southeastern Pennsylvania include approximately 114 stream miles. The system drains approximately 477 square miles and is part of the Chesapeake Bay watershed. The Conestoga River system historically supported breeding and rearing habitat for migratory fish species, including American shad (*Alosa sapidissima*), alewife (*Alosa pseudoharengus*), blueback herring (*Alosa aestivalis*), and American eel (*Anguilla rostrata*). However,

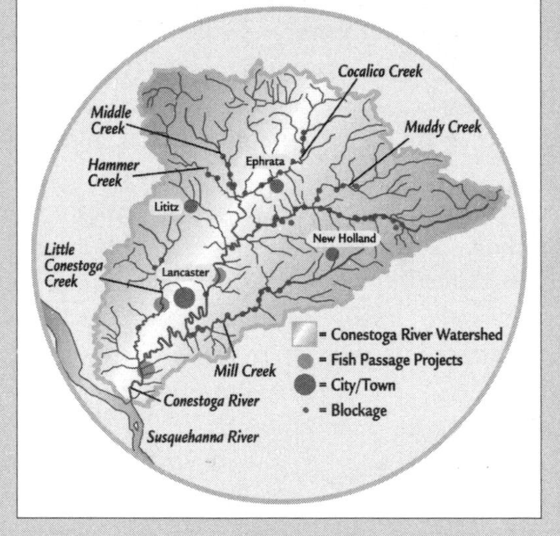

Map of the Conestoga River watershed.

Courtesy of Ted Walke, Pennsylvania Fish and Boat Commission

ogeneity of the substrate, and complexes of useable and available habitats. Water quality criteria include dissolved oxygen levels, thermal characteristics, pH, total suspended sediment, heavy metal concentration, and nutrient levels.

Rivers exist in a dynamic equilibrium and, as a result, the species that inhabit these aquatic ecosystems continuously respond to changes (Vannote et al., 1980). Ultimately, an aquatic ecosystem may not recover to a pre-disturbance condition unless a self-sustaining community based on natural reproduction, succession, and adaptation is attained (Cairns, 1990).

Box 5.4 continued

the 28 artificial blockages (including 23 dams) on the Conestoga River and the 45 artificial blockages (including 44 dams) on its major tributaries made much of the watershed inaccessible to migrating species.

The 1987 Chesapeake Bay Agreement included a commitment that the states that were signatories provide for fish passage at dams and remove stream blockages whenever necessary to restore migratory fish. Since that time, efforts have been undertaken to restore fish passage along the Conestoga River and its tributaries. The Pennsylvania Fish and Boat Commission, through its Consultation and Grant Program for Fish Passage and Habitat Restoration, has been involved in the design and funding of some 20 fish passage projects. In addition, numerous dams have been removed, including four along the Conestoga River (Rock Hill, Eden Paper Mill, Wenger Mill, and Hinkleton Mill dams) and five along its tributaries (Maple Grove, Millport Roller Mill, Lititz Run Intake, East Petersburg Intake, and Martin's dams). All were obsolete run-of-river dams originally built to power mills or supply water to navigation canals. Seven dams removed between 1997 and 1999 varied in height from 3 to 13 feet and in length from 10 to 300 feet, and they cost between $1,500 and $110,000 each to remove (American Rivers, 2001a).

All together, these restoration efforts have reopened more than 28 miles of the Conestoga River to migratory fish. In June 2000, American shad were collected at Lancaster Intake Dam for the first time in decades. In addition to dam removals, efforts are under way to improve water quality in the Conestoga River and its tributaries. Siltation and nutrients have been identified as the two most prevalent causes of quality impairment in the basin. The continued efforts of watershed groups, nonprofit organizations, and government agencies to restore the water quality and integrity of tributary stream channels will have a positive impact and contribute to anadromous fish restoration in the Chesapeake Bay watershed.

Factors Affecting Restoration Rates

The rate of recovery in an aquatic ecosystem after the removal of a dam is difficult to predict due to the large number of controlling factors and the cumulative affects related to integrating ecosystem components (American Rivers et al., 1999; Carmago et al., 1998; Church, 1995; Dadswell, 1996; Iversen et al., 1993). All aquatic systems respond differently to changes and impacts (Table 5.1).

These variables need to be addressed when discussing whether to use an active or passive approach to restoration. The simplest approach is

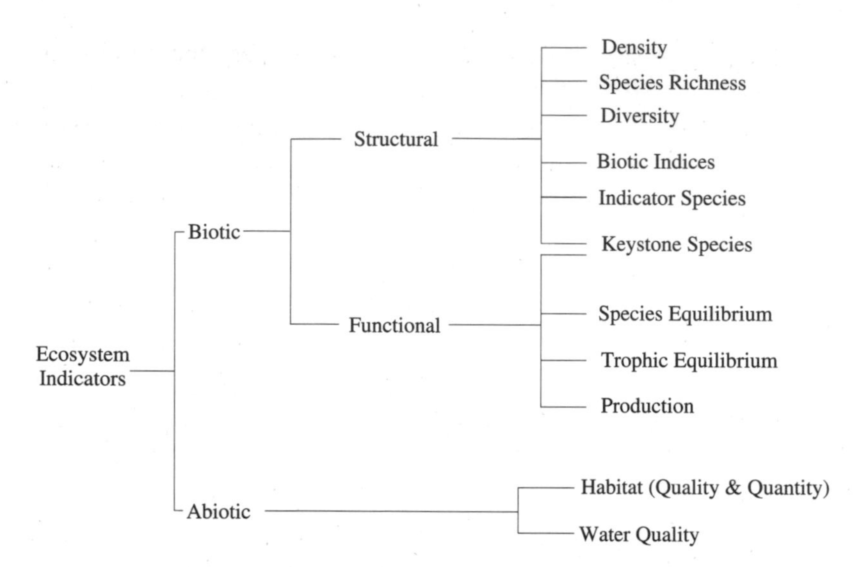

Figure 5.5 A variety of indicators may be used to evaluate aquatic ecosystem recovery. Adapted from Milner, 1994.

a passive one. This approach allows the river system to restore itself with little to no input from stakeholders or managers. The rate of restoration may be considerably slower than that of an active program, and the process may not reach specific objectives and expectations expressed by stakeholders. Small-scale recovery efforts are likely to occur faster than larger-scale efforts. Upstream watershed conditions, riverine inputs, and existing ecosystem dynamics dictate the rate of passive restoration (Iversen et al., 1993; National Research Council, 1992; Nelson and Pajak, 1990; Rabeni and Jacobson, 1993; Shuman, 1995; Staggs, Lyons, and Visser, 1995).

Conversely, an active restoration approach includes collaborators working together and identifying critical ecosystem processes that may need to be jumpstarted to initiate the rehabilitation process. These actions may include the stabilization of sediments, revegetation of exposed sediments, restoration and seeding of specific fish and other species, stabilization of riverbanks, and upstream flow and sediment control. The degree of success depends on how well the restoration group understands the watershed dynamics, timing, finances, personal commitments, and luck required for restoration to occur (Milner, 1994).

Table 5.1 Variables Affecting Rates of Aquatic Ecosystem Response to Dam Removal

Variable	Response
Level of impact (small, medium, or large)	Small systems react more quickly.
Location of dam in watershed	Systems higher in watershed are typically smaller and show faster response to physical restoration.
Hydrologic regime	Higher flow conditions mobilize sediments faster and reset system to new base level.
Time of year	Removal during winter elicits different rates of response than restoration begun in spring.
Expectations of community	High expectations require more active approaches.
Size of watershed	Large watersheds usually have larger repopulation source and consequently may show faster restoration rate.

In many situations, a combination of passive and active approaches provides the best mix and is most acceptable. Stakeholder groups and managers need to work cooperatively to address the issues, identify critical ecosystem relationships, and identify specific actions that will have the highest potential for jumpstarting and guiding the restoration effort.

Physical Habitat

The size of the dam, the reservoirs, and its location in the watershed influences the rate and potential restoration capacity of the river. Dams affect the physical habitat by reducing the amount of sediments downstream, removing the woody debris, causing loss of heterogeneity of the river bed substrate, reducing seasonally dynamic flow patterns, eliminating diversified flow patterns in the river channel, and reducing the heterogeneity of aquatic habitat. These effects influence the colonization times of macroinvertebrates, availability of the suite of aquatic habitats needed by native fish species, and seasonal availability of unique aquatic habitats for spawning and juvenile rearing.

Restoration of Terrestrial and Riparian Vegetation

An aquatic ecosystem is defined by the watershed, the terrestrial, and riparian vegetation that it supports (Williams, Wood, and Dumbeck, 1998). The restoration of the physical and biological components of an aquatic ecosystem depends on the recovery of the riparian corridor along the river (NRC, 1998; Shuman, 1995). The riparian and watershed community provides organic input in the form of carbon necessary for aquatic food production, woody debris for microhabitats, and shade and cover in the aquatic ecosystem (Newcombe and MacDonald, 1991; Nilsson and Dynesius, 1991; Nilsson, Jansson, and Zinko, 1997). A less-disturbed riparian corridor may allow the restoration of the aquatic ecosystem to occur at a faster rate.

Size of Disturbed Area and Upstream Sources of Drift

The size of an aquatic ecosystem and its location in the river's watershed influences the recovery rate (Shuman, 1995; American Rivers et al., 1999). If the area affected by the dam is large, the distances upstream and downstream to likely sources of colonizing macroinvertebrates and fish are likely to be great. Drift from upstream sources and migration of downstream sources are the two primary mechanisms for the natural colonization of the aquatic ecosystem.

Continued Disturbances

Continued disturbances in the form of upstream watershed effects (land use practices) slow or restrict the successional pathways and limit the potential success of a restoration effort. Disturbances may take the form of short-term, limited duration events or longer-term events that affect the whole watershed (Williams et al., 1998). Examples of the latter are increased sedimentation due to logging, mining impacts, upstream water quality impacts, and upstream flow control. Shorter-term disturbances many include localized impacts such as vehicular transport, seasonal livestock movement, and site-specific meteorological events.

Frequency of Previous Disturbances

An aquatic community that historically has experienced frequent disturbances may be restored more quickly than an ecosystem that has a long history of stability (Carmago et al., 1998; Church, 1995; Iversen et al., 1993). Many aquatic ecosystems have a long history of frequent disturbances because of watershed and localized events. In these types of ecosystems, a networked series of metapopulations typically exist in the stream networks that readily supply a new source of colonizing creatures (Peacock et al., 2002). This is analogous to a prairie system that experiences periodic fire. River ecosystems are dynamic; those in diverse environments evolved with the capability to handle varying levels of disturbance. In fact, it is the annual disturbance regime, ranging from floods to low flows, that dictates the productivity of the aquatic ecosystems in many river systems (Allan, 1995).

Presence and Proximity of Refugiums

The farther a refugium is located from the source of recolonizing individuals, the longer it takes for recovery to occur. A natural aquatic ecosystem is composed of a multitude of complex habitats, including microhabitats, migration corridors, unique reaches, riparian vegetation, floodplains, hyporheic zones, production areas, spawning areas, and refugiums (NRC 1992; Shuman, 1995). The refugiums may be in unique areas populated by biota that are sources for the recolonization of rivers following disturbances that reduce biomass or distribution (Sedell et al., 1990). Location of a refugial population necessary for aquatic ecosystem recovery should be identified prior to dam removal.

Flushing Capacity and Persistence of Disturbance

The persistence of the effects of a dam and the sediments in the reservoir has a major influence on the ability and rate of recovery of an aquatic ecosystem. Sediment storage behind dams is often a major issue that needs to be dealt with in any recovery scenario. Sediments in reservoir basins, especially the delta ones, are eroded quickly; these sediments are not extensively consolidated and protected by the roots of riparian vegetation or protective cover (Staggs et al., 1995). Active erosion of these exposed sediments into the river leads to an initial surge of sediment until a level of equilibrium is reached. The erosion of sediments and its movement into the river requires a thorough review of the hydrologics of the river

(Wood and Armitage, 1997). If an aquatic ecosystem experiences periodic (i.e., seasonal) flushing flows, the rate of recovery is enhanced. If, on the other hand, sediments cannot be flushed from the system quickly, recovery is delayed.

Watershed Characteristics and Land Use

Watershed stability and land use influence the rate of recovery, transport of contaminants, and magnitude and frequency of water and sediment disturbances (Williams et al., 1998). Watersheds with extensive and intensive agriculture, logging, mining, or other disturbances have increased amounts of sediment and flashier flow regimes (Williams et al., 1998). Such watersheds are unlikely to reach the objectives for ecosystem recovery after dam removal.

Timing of Disturbance and Life Cycles of the Biota

The time of year when an aquatic ecosystem changes is very important in a determination of the ability and the rate of the species and ecosystem's response (American Rivers et al., 1995; Winter, 1990). The timing of dam removal determines which life stage of an organism is present. Some life stages may be, at crucial times, better able than others to recolonize or survive. The sequence of colonization, succession, and possible end-point community may be influenced significantly by which species are part of the recolonization pool at the time the dam is removed.

Nutrient Input and Recycling

Disturbances that affect autotrophic production or allochthonous inputs to an aquatic ecosystem influence recovery rates of macroinvertebrates and fish. The combined processes of nutrient cycling and transport occur at various rates depending on the productivity of the aquatic ecosystem (Ward and Stanford, 1979; Ligon et al., 1995). The recovery of an aquatic ecosystem following the removal of a dam depends largely on nutrient cycling and retention (Heller et al., 1999; Camargo 1998). In aquatic ecosystems with low nutrient inputs and low turnover and cycling of nutrients, resilience is low and recovery takes longer. Recovery times are reduced if nutrients can be retained in the aquatic system (Kline et al., 1997).

Location of Disturbance in Stream Course and Stream Order

River systems are composites of tributaries and larger rivers. The location of a dam to be removed in a river system has bearing on the timing and restoration potential of the aquatic ecosystem (American Rivers et al., 1999; Iversen et al., 1993). Dam removal lower in the river system offers access to a larger base of recolonization organisms.

Rivers begin a modification process immediately after the removal of a disturbance as flows begin to support the reestablishment of the physical and biological processes that define a dynamic river system (Kinsolving and Bains, 1993). The rate at which rivers modify themselves is a factor of the watershed; flow regimes; time of year; and access to resupply of critical chemical, biological, and physical components (Petts, 1984). Modification or restoration rates are determined by the many factors outlined above, which may be quite variable. Recovery rates associated with dam removal depend on the size of the dam and reservoir located above it, location of the dam in the river basin, upstream watershed disturbances, channel modifications, hydrology, and water quality impacts (Iversen et al., 1993; NRC, 1992; Nelson and Pajak, 1990). In the case of Edwards Dam in Maine, significant signs of a modified river were seen only three months after the dam was removed (Box 5.5).

Water Quality

The water quality of the reservoir basin and upstream watershed can play a very important role in the recovery rate of an aquatic ecosystem. An understanding of watershed dynamics upstream of the restored reservoir and river area is essential to the effective management of dam removal and the prediction of potential for success. Specific water quality components that need to be evaluated include upstream watershed integrity, temperature, dissolved gases, sediment, heavy metal mobilization, and organic matter transport (Murakami and Takeishi, 1997; Newcombe et al., 1991; Shuman, 1995). It is essential that an understanding of watershed dynam-

Box 5.5 Signs of Recovery: The Removal of Edwards Dam in Maine

The removal of Edwards Dam in Maine was followed by a rapid ecosystem response. Edwards Dam was a rock-filled, timber crib structure, 24 feet tall and 917 feet wide, built on the Kennebec River in 1837 for navigation and later used for hydropower generation. In 1993, the dam owner, a small, privately held company, submitted an application to the Federal Energy Regulatory Commission (FERC) to renew its license. A series of studies showed that restoring passage for several migratory fish species would cost 1.7 times more than removing the dam. Moreover, removing the dam would open up 17 miles of historical upstream spawning habitat. FERC denied the dam owner's petition, and in 1999, Edwards Dam was removed.

Signs that the Kennebec River changed in response to the dam removal appeared just a few months later. Bird species such as bald eagles and the great blue heron, once rarely seen, became a common sight along the Kennebec. Less than three months after the removal, schools of striped bass were seen feeding on alewife upstream of the former dam site, and anglers 19 miles upstream caught stripers up to 40 inches long. Populations of 10 migratory fish, once rare in this stretch of the Kennebec, are expected to continue rebounding over the next 20 years (American Rivers et al., 1999).

ics and reservoir basin conditions be developed before initiating dam removal. The timing and management of dam removal needs to take into consideration water quality conditions at the time of removal and how this may affect the watershed.

Upstream Watershed. Decision makers need to understand the watershed dynamics upstream and downstream of the project area, or else the aquatic system recovery may be compromised. Important watershed relationships to evaluate include inflow hydrology and water management; groundwater and sediment supply, quality, and transport; water quality conditions (affected by upstream dams or watershed conditions); riparian well-being; upstream aquatic assemblages (source for replenishment); and upstream land use (intact or fragmented) (Williams et al., 1998). If the upstream integrity of the watershed is high, then the aquatic restoration probably will succeed. If the upstream watershed is heavily affected, the potential for success may be reduced significantly.

Temperature. As reservoirs are drained, the reservoir water body is transformed back to the conditions of the river system. Fish and insect species that have adapted to reservoir conditions are likely to be affected as dynamic river conditions begin to be reestablished. Thermal regimes defined by the upstream watershed, groundwater, and seasonal ambient conditions become a primary factor defining the aquatic assemblage (Vinson, 2001; William and Hynes, 1976). Specific evaluations need to be made of the presence and impact of upstream dams and reservoirs and the potential thermal regime supplied to the recovery area.

Sediment. As reservoirs are drained, sediment deposits trapped behind the dam are subjected to hydraulic forces. Local mobilization of sediments occurs as a reservoir drops in elevation and the inflowing river begins to migrate across and through the deposited sediment. Initially the delta area will be eroded and the process will continue downstream (Shuman, 1995; Dadswell, 1996; Department of the Interior, 1996). The mobilization of sediment is likely to lead to increased sediment and turbidity levels in the reservoir basin and immediately downstream of the removed dam. This increased turbidity and sediment transport may affect the eggs of fish species that have been deposited in cobbles and gravels downstream, and insect species that depend on clear water conditions (Kondolf et al., 1993; DOI, 1996; Newcombe and MacDonald, 1991). The level of impact depends on the season of drawdown, the rate of reservoir drawdown, and the upstream watershed dynamics. The potential impacts related to sediments are large and need to be evaluated carefully in terms of timing and the management of dam removal.

Heavy Metal Mobilization. Reservoirs located in watersheds where mining occurs or historically has occurred may accumulate mining waste and heavy metals, which may become trapped in the sediments behind the dam (Murakami et al., 1997). The sediment composition and quality, and the potential for remobilization and transformation of the metals and mining wastes, need to be considered before dam removal is initiated. The rate of timing of the reservoir drawdown combined with on-site stabilization and upstream watershed dynamics may be necessary to avoid water quality impacts downstream (DOI 1996; American Rivers et al., 1999).

Dissolved Gas. Dissolved gases occur naturally in all water bodies. Supersaturation of the water column with dissolved gases can result from both natural and human-induced conditions. Dams often exacerbate the effect of dissolved gases on the downstream aquatic environment. Water released through dams may increase the levels of total dissolved gases in the water column and negatively affect the health and survival of young-of-the-year, juvenile and adult salmonids, and aquatic insects (Weltkamp and Katz, 1980; NRC, 1992; Ligon et. al., 1995). Careful timing and management of dam removal and subsequent water releases are essential to avoid unnecessarily modifying the water quality and affecting the downstream aquatic biota.

Organic Matter Transport. Organic matter in the form of materials that have been trapped in reservoir basins may be mobilized as dams are removed and water bodies drained. Specifically, sunken trees, wastewater treatment residue, aquatic plants, and terrestrial vegetation may be mobilized and may generate a spike in organic matter and carbon supply downstream of the removed dam (Shuman, 1995; DOI, 1996; Dadswell, 1996).

CONCLUSION AND RECOMMENDATIONS

One way to learn about the potential effects of dam removal is to review what is known about the effects of dam installation on a river system. Although the changes brought about by installation may not be completely reversible, they do help predict the various consequences of removal. Changes in the physical system of a river imposed by a dam, and partly reversed by dam removal, cause associated adjustments in the biological components of the ecosystem. These biological changes, particularly among fish and macroinvertebrates, include altered movement patterns, residence times, species assemblage, and general habitat opportunities. These biological ecosystem changes are variable in time and space. The extent and intensity of the changes depend on the size of the dam (i.e., storage capacity), quantity and quality of sediment in the reservoir, timing of reservoir level fluctuations, limnological conditions in the reservoir, and stability of the downstream river reach. Non-native exotic species also affect native species in both rivers and reservoirs.

Dam removal has increased the abundance and diversity of aquatic insect, fish, and other populations, but long-term data and numerous "before and after" tests of population trends are not available.

Reservoirs create wetland areas in some cases; the removal of a dam and draining of a reservoir may create some wetlands downstream, but at the expense of some wetlands upstream. Dam removal often results in the replacement of one aquatic community with another that is partly natural and partly artificial. The most significant biological effect of the removal of small structures is the removal of physical obstructions and increased accessibility of upstream habitat and spawning areas for migratory fishes.

- **Conclusion:** Decisions to remove dams have far-reaching implications both upstream and downstream in a complicated physical and biological system. The consideration of a limited scope of outcomes is likely to have unforeseen consequences.
- **Recommendation:** The panel recommends that dam removal decisions take into account watershed and ecosystem perspectives as well as river reach perspectives and the more limited focus on the dam site.
- **Recommendation:** The panel recommends that the U.S. EPA and/or appropriate state or local government agencies conduct a monitoring and evaluation program following dam removal. This program should be developed and implanted so that vital data on the natural and enhanced restoration of habitats is collected and made available in public datasets for use in adaptive management.

6

Economics and Dam Removal

Dam removal is not unambiguously good, but attaching a more precise valuation is difficult because formal benefit–cost analysis procedures do not necessarily apply to dam removals. Existing procedures are intended for the evaluation of federal water resource development projects, especially the construction of large dams. Even if a particular dam removal qualifies as a federal action, economic analysis is secondary to certain overriding environmental considerations, such as the preservation of endangered species, or to safety concerns.

On the other hand, few dam removals are without controversy, and most involve numerous and diverse stakeholders, many of whom may be concerned about the justification for the ultimate decision. Benefit–cost analysis provides a disciplined process for identifying and measuring all potential effects of removal, both positive and negative. It arrays the impacts on various stakeholders in a way that allows comparisons to be made, and trade-offs negotiated.

The application of the federal benefit–cost analysis paradigm to proposed dam removals produces an interesting and somewhat complex inversion of the issues familiar to dam builders. On first impression, it may appear that the beneficial and adverse effects of dam construction would simply change sides. That is, the beneficial effects of dam removal might be thought of as the avoided costs of dam operation and avoided external costs, and the adverse effects of dam removal might be the lost beneficial effects of dam operation.

Although some of these relationships may be relevant, a number of new issues intrude on dam removal economics. These include the problem of defining a reference case. Conventional benefit–cost analysis typically defines a no-action alternative, in comparison to which all beneficial and adverse effects are measured. With an existing dam, possibly in deteriorated condition and perhaps unsafe, no action may not be an alternative. A different type of reference case needs to be specified. For example, in cases in which it can be assumed that the monetary benefits of removal (avoided rehabilitation and operating costs) exceed the monetary costs (dam removal costs and the value of lost services), the non-quantitative environmental benefits might not be considered. This does not mean they are unimportant. In fact, states may be supportive precisely because dam removal assists them in achieving policy goals of improved water quality and habitat, and availability of fish and game species.

Dam removal may serve multiple economic objectives. The removal of Hinkletown Dam in Pennsylvania, for instance, restored fish habitat and cleared the way for new bridge construction (Box 6.1). When a dam is removed in the hope of restoring fisheries and/or various riparian environments, resources that were lost when the dam was built are not necessarily the ones that will be recovered. Future fish runs may differ from past ones for various reasons, riparian vegetation may regrow in different ways from its historical condition, and stream morphology may change as well. Introduced or exotic species in the area or reservoir also may compromise restoration goals. To prepare a credible economic analysis, the analyst not only needs to predict what will happen, but also needs to say when. In some cases there may be a significant lag after dam removal before ecosystem restoration or other removal objective is realized or attains management objectives. Because of discounting in the economic assessment, such lags can greatly reduce the weight given to a particular beneficial effect.

The adverse effects of removal include a number of straightforward cost items, such as the cost of removing the structure and disposing of the debris. However, this category also includes costs for which little data or expertise exists. For example, it is necessary to predict how the sediment will move after the dam is breached, and to identify adverse environmental impacts associated with the sediment load. The movement of clean sediment can be beneficial to downstream beaches (e.g., the case of Rindge Dam). Furthermore, the duration of these impacts also needs to be estimated. As with certain items on the benefits side of the ledger, there is only modest experience with this phenomenon, and estimates of magnitude and duration of dam removal effects necessarily involve substantial uncertainty.

Box 6.1 The Removal of Hinkletown Dam in Pennsylvania

Dams may be removed for a variety of reasons, including those unrelated to safety, species management, or river restoration. Hinkletown Mill Dam was a rock-filled timber crib dam capped with concrete located 38.7 miles above the mouth of the Conestoga River in Pennsylvania. The dam was approximately 7 feet high with a crest length of 92 feet. It originally was built in the 1700s to provide waterpower for a flourmill, which ceased operations sometime between 1940 and 1965. In 2000, the state highway department removed this obsolete, run-of-river dam. The realignment of approach roadways and modifications to a state highway bridge required the construction of new bridge piers at the same location as the dam.

Hinkletown Mill Dam before removal (upper photograph), and the river during the removal.

Photos courtesy of the Pennsylvania Department of Environmental Protection

NO-ACTION ALTERNATIVE

The analyst needs to define a point of reference for use in identifying and measuring the beneficial and adverse effects resulting from any action. Conventionally, this has been a no-action alternative: a scenario of events characterized by the absence of the action under study. The consequences of the proposed action are identified by comparing present and future conditions with the action to present and future conditions without the action. For example, if it is argued that a dam to be built will provide downstream flood protection, it needs to be shown that the expected flood damages with the dam (proposed action) will be lower than those expected without the dam (no action). The same "with–without" logic applies to all beneficial and adverse effects of a project.

However, the term "no action" is something of a misnomer. In almost every case, it involves action of some type. Conventional benefit–cost analysis requires that the no-action alternative be both feasible and the most likely set of events in the absence of the action under study. For example, if a water supply dam is planned to replace a community's contaminated groundwater, no-action alternatives do not include the continued supply of contaminated water or the absence of water supply. Instead, the analyst needs to determine the community's most likely response to the water supply problem if the dam is not built (e.g., importing water from another community or drilling new wells into a different aquifer). The no-action alternative incorporates whichever strategy proves feasible and is found to be the most likely choice. Benefit–cost analysis of the proposed dam considers only the differences between outcomes if the dam is built versus if it is not (and the no-action plan is implemented).

This same logic can be applied to dam removal. Many dams proposed for removal have structural or other safety deficiencies and some may no longer serve the purpose for which they were built. Furthermore, these conditions can be interdependent. Deteriorating turbines and generators may lead to the abandonment of a privately owned hydroelectric dam, especially if the combined cost of powerhouse renovation and needed (safety-related) structural improvements to the dam makes further investment infeasible. Conversely, if a dam no longer serves any economic purpose for some external reason, there may be a reduced willingness to carry out safety-related improvements.

In defining a no-action alternative to a proposed dam removal, it is clear that "no action" needs to include whatever actions are necessary to protect human life and comply with applicable regulations. These actions may include rehabilitation of the structure, spillway enhancements, and appropriate maintenance of the dam over the planning period. In this way, the avoided life-cycle cost of the safety upgrade becomes a beneficial effect of removal. But it is also possible that a previously abandoned dam (one that no longer serves any economic purpose) will become a viable asset once again after the safety-related deficiencies are corrected.

The no-action option also needs to take into consideration the anticipated costs, direct and indirect, that will be incurred if actions to protect human life fail. Because many small dams are privately owned, the loss of life due to dam failure has extenuating social aspects that, although difficult to measure, may have a significant impact on the dam owner and surrounding community.

VALUING THE OUTCOMES OF DAM REMOVAL

Market versus Nonmarket Goods

Some dam services, such as hydroelectric energy, are market goods. These services are sold or can be sold in existing markets at prevailing prices. The market determines values, and market transactions provide the data necessary for calculation. Certain outcomes of dam removal, such as the cost of breaching or the cost of removing portions of the structure, are also market goods. In this case, market transactions provide information on the value of labor, materials, machines, and so on used in the effort.

However, many outcomes of dam removal are not market goods and cannot be valued directly using market data. These outcomes include lost dam services, such as recreation, irrigation, water supply, and flood protection. These are nonmarket goods because users are not asked to pay any price (e.g., for recreation, flood protection), or the price is set administratively and does not reflect any market phenomena (e.g., for irrigation water). The environmental changes produced by dam removal (including the restoration of fish habitats) are also nonmarket goods because they are not priced and have no near substitutes that can be valued as market goods.

Methods are available for placing monetary values on nonmarket goods in some, but not all, circumstances. The commonly used methods can be organized into two general approaches. The revealed preference approach includes methods requiring that the nonmarket good be a weak complement for some market good. In this case, the characteristics of the market good can be used to impute the value of the nonmarket good. Certain other nonmarket goods, not necessarily related to market goods, can be valued directly using one of a number of stated preference methods. Freeman (1993) is a standard reference for nonmarket valuation methods.

Revealed Preference Valuation Approaches

In the water resource field, two revealed preference methods have been widely applied: the travel cost method and hedonic price analysis.

- **Travel Cost Method.** Some services provided by dams, such as water-based recreation, must be used *in situ*. No one can use these services unless they travel to the location of the dam and reservoir. Even though the recreation service may not be priced, users reveal something of their valuation for that service through their willingness to incur travel costs (the weakly complementary market good). The application of this method usually involves the recording of automobile license numbers at recreation sites as well as detailed interviews with a sample of visitors. Certain aspects of the method are controversial, such as the usual assumption that the only purpose of a respondent's trip was to visit the site in question. If other stops were made, or if the trip itself was considered pleasurable, the travel cost method may overestimate the value of the service. On the other hand, other factors may cause the method to underestimate the value.
- **Hedonic Price Analysis.** If the visual amenity of a lake benefits property owners near the shore of the lake, then the amenity can be said to be consumed as part of a bundle of market and nonmarket goods (housing services, location convenience, etc.). Because the value of the amenity is assumed to decline on a smooth gradient as properties are located farther from the shore, it is possible to design a statistical analysis of many property values that will separate the component of property price attributable to the amenity. This method is quite limited in application because it can deal only with nonmarket goods that are bundled with market goods, and it further requires rather large datasets and carefully executed

statistical analysis. Biases are also possible, especially when the analysis does not satisfactorily control for housing attributes that may be correlated with distance from a shoreline.

There are additional market-based methods, including methods based on avoidance costs, or alternative costs. These are not included here because they are not widely applied to water resource projects.

Stated Preference Valuation Approaches

Stated preference methods are also described as direct valuation methods because they do not rely on related goods or actual markets in any way. Rather, these methods solicit valuations directly from users. The most common stated preference methods rely on survey research, most often in the form of personal or telephone interviews, or mailed questionnaires. The survey instruments describe a hypothetical market for the nonmarket good and ask respondents to state their valuations in one of a number of ways.

- **Contingent Valuation Method.** This is the most familiar stated preference approach. At the simplest level, the nonmarket good is described, a hypothetical market transaction is proposed, and the respondent is asked what he or she would pay for the good (or what payment would be accepted to forgo the good). The actual question may be open-ended ("What would you pay?"), multipart, or based on a payment card or some other device. The survey also includes questions on personal attributes (age, gender, education, income, etc.) that are used later to extrapolate sample responses to a larger population.
- **Contingent Referendum Method.** This is similar to the contingent valuation approach in all respects except that the elicitation question has the form "Would you pay $__, yes or no?" This is generally an easier question for respondents to answer, and it avoids some types of bias, but the method requires a much larger sample to produce useful results.
- **Contingent Ranking Method.** With this method, the respondent is not asked for a monetary value but instead is asked to rank a number of situations that involve different levels of the nonmarket good in question. This establishes a type of value relative to the (possibly known) values of other goods involved in the

ranked alternatives. With careful design, this method can produce a monetary valuation.

- **Contingent Activity Method.** The respondent is asked how he or she would vary certain activities in response to a gain or loss in the nonmarket good. For example, if a dam is to be removed, nearby residents might be asked how often they would travel to the site, as opposed to their travel habits before the removal. An examination of the activities may provide a basis for valuing certain nonmarket goods.

All stated preference methods depend on the skill with which the survey instrument is designed and applied, as well as the sample size and the way in which results are analyzed. In addition to a possibly large error arising from the hypothetical nature of the valuation, there are numerous sources of potential bias. These include sample bias, nonresponse bias, strategic bias, starting point bias, anchoring, and implied value cues. Properly designed instruments administered by well-trained, professional interviewers generally avoid these biases or reduce them to low levels. The inherent error (sometimes called hypothetical bias) depends, in part, on the type of good being valued. If respondents have experience with purchases of a similar good or otherwise can imagine that good being traded in a market, the hypothetical may be minimal. If respondents cannot conceive of an actual market transaction involving the same or a similar good, then the questions may be difficult to answer and the hypothetical bias large.

Following the widely noted use of contingent valuation (CV) to estimate damages resulting from the *Exxon Valdez* oil spill, concerns were raised in the literature regarding the economic consistency of CV results, especially in the presence of significant non-use values. In response, the National Oceanic and Atmospheric Administration (NOAA) convened a panel of prominent social scientists, co-chaired by Nobel laureates Kenneth Arrow and Robert Solow. The NOAA panel developed a comprehensive set of guidelines designed to ensure reliable CV studies (*Federal Register*, Volume 58, pp. 4601–4614 [1993]). The panel's principal addition to the previous literature on this subject was the subset of guidelines described as the "burden of proof" requirements. These include criteria such as an acceptably low nonresponse rate, responsiveness of valuation to scope of damage, and respondents' understanding of the task. A later review by Carson et al. (1996) demonstrated that CV studies using the best practice of the time, including the *Exxon Valdez* study, complied fully with the NOAA panel guidelines. With respect to professionally designed and executed studies, the review concluded, "the Panel's concerns about temporal reliability, question format, and social desirability biases appear unwarranted" (Carson et al., 1996).

Generally, stated preference methods are broadly applicable to a variety of nonmarket goods. They also have the unique capability to measure intrinsic values (existence value, option value, bequest value) as well as use value. Properly done, the valuations produced by these methods can be credible and reasonably accurate. However, the techniques involved are very demanding. The necessary work is costly and time consuming and requires a high level of skill and experience. Anything less runs the risk of producing severely biased results.

BENEFICIAL OUTCOMES OF REMOVAL

Restored Environmental Services

The removal of a dam of almost any size usually has a profound effect on the stream and its riparian environment. Specifically, the stream flows freely again; there is no longer a distinction between upstream and downstream areas in the reach containing the dam site. Land previously inundated is exposed and revegetated. Slack water habitats and flat-water recreation areas may be lost, and stream habitats may be expanded and reconnected. Some fish habitats are lost, and others re-created. Although many dam removal decisions may be prompted by issues of human safety or other potential hazards, increasingly the restoration of fish habitats and fluvial processes also motivate dam removals, especially where dams have functioned as migration barriers to spawning by anadromous fish populations.

However, it is important to note, as emphasized in the previous chapter, that the restored habitats and biological communities will not necessarily be identical to those that were lost when the dam was constructed. Fish runs may or may not approximate those of historical record and may develop only after some time. Exposed land may revegetate with exotic trees or plants. An assessment of restored environmental functions, therefore, requires a determination of what is likely to be created and how long it will take.

The economic consequences of restored environmental functions related to dam removal are of two types: use values and intrinsic values.

Use values are economic measures of valuable environmental services that result from environmental functions. For example, the recovery and expansion of fish habitat (function) may lead to an increased population of harvestable fish (service). This increase can be quantified by a comparison to populations associated with the no-action alternative. Then the increased population can be converted to increased catch by commercial fishers and/or increased recreational fishing days, as appropriate. These predicted outcomes are economic goods, which can be valued by any one of a number of methods, providing a monetary measure of the use value of restored environmental services.

Intrinsic values are not directly related to the economic use of a dam or reservoir. Some restorations of environmental functions are regarded by society as valuable in their own right, simply because of their existence or the knowledge that resources will be preserved for future generations. These intrinsic values are also known as option values, existence values, and bequest values. They are likely to appear in cases in which the affected resource functions are unique in some way (no similar functions are generally available in other places) and the action that creates or destroys them is essentially irreversible. Irreversibility often is defined to include cases in which a reversal of the action is possible but very unlikely on economic grounds. The only available methods for placing monetary values on intrinsic values are the family of stated preference methods: contingent valuation, contingent referendum, or factorial analysis (Freeman, 1993; Randall, 1991). The use of these valuation methods requires substantial skill on the part of the analyst and may be costly or infeasible in particular situations.

Avoided Costs

When the costs associated with dam removal are compared to the costs implied by a properly designed no-action alternative, it becomes apparent that most direct costs of the no-action alternative are avoided by dam removal. These include the costs of rehabilitating the existing structure, making any required enhancements (spillway reconstruction), and maintaining the existing structure throughout the planning period, including liability insurance costs. These avoided costs are among the beneficial outcomes of dam removal. Sometimes the avoided costs of rehabilitating the structure drive the removal decision, as in the case of the Sandy River dams in Oregon (Box 6.2).

Box 6.2 Removal versus Renovation Costs: The Case of the Sandy River Dams in Oregon

Two dams on the Sandy River in Oregon (Little Sandy Diversion Dam and the Marmot Dam) will be removed soon because the costs of the renovations necessary for renewal of their Federal Energy Regulatory Commission (FERC) license would be much greater than the costs of removal. Marmot Dam, built in 1912, is a concrete gravity dam that is 47 feet high and 195 feet long. Its reservoir area is filled completely with sediment and unusable for water storage. Little Sandy Diversion Dam, built in 1906, is smaller, with a height of 16 feet and a length of 114 feet. Portland General Electric (PGE) currently owns and operates the two dams under a FERC license. Marmot Dam and Little Sandy Dam divert water to the Bull Run Hydroelectric Project, which generates 22 megawatts of electric power. The FERC license for the hydroelectric project will expire on November 16, 2004.

On May 26, 1999, PGE announced its decision to surrender its operating license and decommission the project because the two dams would need costly renovations before the license could be renewed. The estimated cost for removing the two dams is $22 million and will be covered by PGE, the city of Portland, and the state of Oregon (Environmental News Network, 1999). Portland, the first major urban area to have a fish listed under the federal Endangered Species Act (ESA), is 30 miles west of the two dams and is interested in this project because it helps the city comply with ESA requirements (Environmental News Network, 1999). The removal of Little Sandy Dam would open up a 12-mile stretch (now virtually dry) of the Little Sandy River Basin to salmon and steelhead trout. The removal of Marmot Dam would open up 10 miles of the Sandy River. The National Marine Fisheries Service and other federal and state agencies also will provide assistance with this project.

ADVERSE OUTCOMES OF REMOVAL

DIRECT COSTS OF REMOVAL

As in the case of avoided costs, a comparison of dam removal costs to those implied by the no-action alternative identifies a number of direct costs incurred only in the case of removal. These include the cost of studies and investigations needed to plan the removal effort, monitoring costs, the cost of breaching and removing the physical structure, and the cost of managing sediment flows. Table 6.1 shows a sampling, grouped by state, of the total costs of various dam removals in the United States.

Table 6.1 Examples of Dam Removal Costs in the United States[a]

State	Watercourse	Project Name	Height (feet)	Length (feet)	Removal Costs ($ million)
CA	Cold Creek	Lake Christopher Dam	10	400	0.100
CA	Lost Man Creek	Upper Dam	7	57	0.029
CO	Ouzel Creek	Bluebird Dam	56	200	1.500
FL	Chipola River	Dead Lakes Dam	18	787	0.032
ID	Colburn Creek	Colburn Mill Pond Dam	12	35	0.030
ID	Clearwater River	Lewiston Dam	45	1,060	0.633
ME	Kennebec River	Edwards Dam	24	917	2.1
ME	Pleasant River	Brownville Dam	12	300	0.078
ME	Pleasant River	Columbia Falls Dam	9	350	0.030
ME	Souadabscook Stream	Grist Mill Dam	14	75	0.056
ME	Stetson Stream	Archer's Mill Dam	12	50	0.013
MI	Muskegon River	Newaygo Dam			1.300
MN	Cannon River	Welch Dam	9	120	0.046
MN	Kettle River	Sandstone Dam	20	150	0.208
NC	Little River	Cherry Hospital Dam	7	135	0.069
NC	Neuse River	Quaker Neck Dam	7	260	0.206
NM	Santa Fe River	Two-Mile Dam	85	720	3.200
OH	Little Miami River	Jacoby Road Dam	8	100	0.010
OR	Bear Creek	Jackson Street Dam	11	120	1.200
OR	Evans Creek	Alphonso Dam	10	56	0.055
PA	Conestoga River	American Paper Products Dam	4	130	0.060
PA	Conestoga River	Rock Hill Dam	13	300	0.110
PA	Little Conestoga River	East Petersburg Authority Dam	4	20	0.005
PA	Little Conestoga River	Maple Grove Dam	6	60	0.017
PA	Muddy Creek	Amish Dam	3	40	0.002
PA	Muddy Creek	Castle Fin Dam	5	383	0.210
VT	Clyde River	Newport No. 11 Dam	19	90	0.550
WA	Whitestone Creek	Rat Dam Lake	32	240	0.052
WI	Baraboo River	Waterworks Dam	14	220	0.213
WI	Bark River	Slabtown Dam	10	60	0.030
WI	Eighteen Mile Creek	Colfax Dam	20	350	0.241
WI	Manitowoc River	Manitowoc Rapids Dam	16	400	0.045
WI	Milwaukee River	Milwaukee Dam	19	432	0.345
WI	Pine River	Parfrey Glen Dam	19	450	0.154
WI	Willow River	Mounds Dam	58	430	0.170
WI	Willow River	Willow Falls Dam	60	160	0.450

Source: Data from American Rivers et al. (1999).

[a] These costs are for dam removal only and do not include site, reservoir, or downstream restoration.

Lost Dam Services

Dams typically provide a number of services, even when they are intended as single-purpose assets. The impoundment behind a flood control dam also may support recreational uses, enhance the market value of surrounding property, and provide valued fish habitat for introduced game species. A water supply reservoir may offer flood protection benefits for downstream property. When dam services are considered, it is important that the analysis extend to all services that would be supported by the no-action alternative, whether part of the original dam purpose or not. Generally speaking, a dam removal terminates all dam services. In this case, the monetary value of the services associated with the no-action alternative becomes one of the adverse outcomes of removal. Sometimes, blocking fish migration (as a dam might) is seen as beneficial, such as for the protection of upstream habitat for exclusive use by resident species such as bull and rainbow trout, or as a barrier to invasions of exotic species.

External Costs of Removal

Just as dam construction imposes costs on third parties and on society as a whole (environmental costs, typically), dam removal creates external costs as well. These include certain environmental costs associated with the removal itself, such as the temporary loss or degradation of downstream habitat due to sediment flows. External costs also may include the loss of visual amenities, if either the impoundment or the dam structure itself is regarded as a point of interest in the local landscape. For this reason, some dam removal proposals contemplate the preservation of a large portion of the dam structure, to retain some of the visual interest. It is important to conduct a wide search for external outcomes of both dam removal and the no-action alternative, so that a comparison between the two sets of outcomes can reveal the external consequences of removal (Box 6.3).

CHALLENGES FOR ECONOMIC ANALYSIS OF DAM REMOVALS

The application of conventional methods of benefit–cost analysis to dam removals can assist with decision making related to dam removal projects. To some extent, an economic analysis of these projects involves the same

Box 6.3 Matilija Dam: Factors To Consider in Benefit–Cost Analysis

The case of Matilija Dam illustrates the complexities of a benefit–cost analysis of a potential dam removal. Matilija Dam was constructed in 1947 on Matilija Creek, a tributary of the Ventura River in Southern California, to control flood surges and provide a constant supply of water to the Ojai Valley (U.S. Bureau of Reclamation, 2000). The structure is a variable-radius concrete arch dam that stands 190 feet tall and 620 feet wide. Notches were cut in the 1960s to prevent the structure from collapsing and in 2000 for a dam demolition demonstration project. Matilija Dam no longer provides any significant flood control or water storage capacity. Moreover, the structure has blocked endangered steelhead trout from approximately 85 percent of their habitat on the creek and trapped much of the sediment needed to replenish downstream Ventura County beaches (U.S. Bureau of Reclamation, 2000).

Matilija Dam in 2001.

Photo courtesy of Sarah Baish

Today, there is a broad consensus that Matilija Dam's negative impacts greatly outweigh its benefits, and it soon may become the largest dam ever removed in the United States. In the summer of 2001, the U.S. Army Corps of Engineers took the lead in a feasibility study to determine the preferred method for removing the dam and assess the costs and benefits of removal (Ventura County Flood Control District, 2001). The Matilija Dam Ecosystem Restoration Feasibility Study is a collaborative effort that includes many federal, state, and local government agencies and several nonprofit organizations. The study is expected to be completed in 2004 and to cost $4.2 million. The feasibility study needs to consider the following factors in the analysis.

Costs of Removing Matilija Dam

Direct Costs

A 2000 Bureau of Reclamation study estimated that removing the dam would cost $21–$180 million. The wide range in cost is due to

Box 6.3 continued

different methods of dismantling the dam. The least expensive method, but also the one that poses the highest risk of downstream flooding, would be to remove the dam gradually and allow the natural river flood flows to transport the sediment downstream. The most expensive method would be to remove the sediment using a slurry pipeline, depositing it directly on Ventura County beaches (U.S. Bureau of Reclamation, 2000).

Lost Dam Services

- *Potable water supply* The Casitas Municipal Water District would lose approximately 400 acre-feet of water per year, which currently serves 1,000 nearby residents. If the dam is not removed, the reservoir is expected to be filled completely with sediment by 2010 (U.S. Bureau of Reclamation, 2000).
- *Fire- fighting water supply* The reservoir is used occasionally as a water source by fire-fighting helicopters. Alternatives could include nearby Casitas reservoir or the Pacific Ocean.

External Costs of Removal

- *Loss of reservoir wetlands* Approximately 20 acres of wetlands exist at the reservoir, supporting numerous plant and animal species (U.S. Bureau of Reclamation, 2000). Although new wetlands would be created once the dam is removed and the reservoir drained, removal would result in an overall loss in wetlands.
- *Increased sedimentation to downstream dam* Robles Diversion Dam is located downstream of Matilija Dam and currently needs to be cleared of sediment every 5 years. This cycle will increase with the removal of Matilija. A facility that will allow sediment to be flushed through the Robles dam during high flows is being planned (U.S. Bureau of Reclamation, 2000).

Benefits of Removing Matilija Dam

Restored Environmental Services

- *Fisheries* Removing Matilija Dam would open up 30 miles of stream for anadromous species of fish, including 85 percent of the remaining habitat of endangered steelhead trout. The population of steelhead trout has been reduced to fewer than 200 from a historical run of at least 4,000 adult fish per year (Capelli, 1999). One study has shown that a single steelhead may be worth $75 to $300 because of increased sport fishing business revenues (e.g., from fishing and outdoor equipment, lodging, guide services, and restaurant meals). Increasing the sport catch in the Ventura River by 2,000 adult fish (about half of the historical run) could generate as much as $600,000 per year to those industries (Marx, 1996–1997).
- *Beach sediment* Matilija Dam traps much of the natural supply of sediment for replenishing Ventura County beaches 16 miles

Box 6.3 continued

downstream. Estimates show that up to 70 percent of the 50 years of sediment trapped behind the dam is suitable for placement on beaches, an amount sufficient to widen all south county beaches by 30 feet (Marx, 1996–97). Removal of the dam may increase coastal tourism on the beaches, which in 1992 brought in an estimated $45 million to Ventura County (State of California, 1997), and will increase the protection of shorefront property from erosion and storms.

- *Recreation* The dam's removal would increase recreational opportunities within the former dam site and Matilija Canyon. Public access to nearby Los Padres National Forest also would be enhanced.

Avoided Costs

- *Maintenance* The removal would eliminate the cost of constant dam maintenance for the owner, the Ventura County Flood Control District.
- *Dam safety liabilities* The removal would eliminate future dam safety liabilities for the Ventura County Flood Control District.
- *Beach replenishment and other protective measures* To deal with high erosion rates attributable to the dam, costly measures such as beach nourishment, groins, revetments, and a seawall have been used in Ventura County. These structures are falling into disrepair, and multimillion-dollar projects are necessary to maintain them.

types of engineering cost estimates that are familiar to dam builders. The evaluation of one category of dam removal outcomes—lost dam services—may be facilitated if there are usable data on some or all of these services in past years.

Dam removals also have a number of impacts that are not familiar to dam builders. In some cases, the environmental outcomes may be difficult to predict because of a lack of experience with similar events. Predictions of many dam removal outcomes are likely to be quite uncertain, as are the predictions of the times at which such outcomes will appear. Because the anticipated benefits of a removal may consist largely of uncertain environmental changes expected to arrive at uncertain times, this lack of solid information influences dam removal decisions. Furthermore, all outcomes are identified and measured by comparison to a reference case—the no-action alternative. As noted above, the specification of such an alternative raises a number of questions as to what is, and what is not, to be included.

An additional unfamiliar issue concerns the valuation or quantification of various beneficial and adverse outcomes, especially the environmental outcomes noted above. Although such identification and quantification can lead to important and useful data, full participation in the benefit–cost analysis process is granted only to those outcomes that can be described in monetary terms. Although a variety of methods exist for attaching monetary value to environmental resource services, they are relatively costly, require great skill to apply, are applicable only to certain services or under certain conditions, and are characterized by substantial uncertainty.

CONCLUSION AND RECOMMENDATION

Formal economic analyses can be very helpful in supporting the decision-making process for dam removal, in setting priorities, and in considering the interests of stakeholders and agencies. Nevertheless, significant challenges remain for those who would use methods such as benefit–cost analysis for this purpose. Dam removal has various environmental outcomes, including some that are highly uncertain and difficult or impossible to value monetarily. It may be tempting to ignore these issues, as often was done in the earlier building of dams. However, these nonquantified environmental effects are major issues when dealing with removal and need to be taken into account. Rather, it is necessary to confront these dam removal impacts and the challenge of accounting for them, and to focus on developing credible projects and improved methods with which to evaluate them.

- **Conclusion:** The science of economics does not offer decision makers considering dam removal a sufficient array of analytical tools and supporting data to assess adequately the economic outcomes of a decision in quantitative terms.
- **Recommendation**: The panel recommends that the community of economics researchers provide (1) improved economics evaluation tools for dam removals to enable the assignment of monetary valuations for outcomes of dam removal and (2) empirical research on changes in property values associated with dam removals already accomplished.

7

Social Aspects of Dam Removal

Dam removal decisions involve social and cultural issues, which frequently can be contentious. Some of these issues relate to visual aesthetics, recreation, and cultural and historical preservation and/or restoration. Restoration most often is thought of in dam removal processes as an environmental concept, but it also has human social dimensions. The historical restoration that may be a result of dam removal can be a desirable outcome from a community standpoint because of commonly shared social values that revere the area's human history. Social and historical values are important considerations in decisions to remove dams because society in general pays the costs, not only in monetary terms when public funds are involved, but also in terms of the lifestyles experienced by residents of the area. The general appearance of the river after a dam is removed may be important to local residents for purely aesthetic reasons, and the ambience of the river may have far-reaching commercial implications. This chapter reviews some general ideas about social values and aesthetics related to rivers and dams. It also reviews the values of tribal nations in the United States and the use of social impact assessments.

AESTHETICS AND SOCIAL VALUES

Like all natural resources, rivers have use and non-use values (Hanna and Jentoft, 1996). Use values include direct use, indirect use, and option values (see Chapter 6). The predominant U.S. perspective on environmental resources, including rivers and water, has been one of exploitation for economic development. Throughout most of this nation's history, the dominant perspective (anthropocentric) has been to see nature as existing to serve humankind, and a guiding principle in the use of natural resources has been the alteration of natural environments to enhance the wealth and comfort of people. An important ancillary concept is that of private property rights, which is firmly enshrined in the policy and laws of the nation and is a mainstay of the orderly use of resources, including rivers.

Social values for the natural environment, including rivers, have undergone considerable change during the history of the United States. When Europeans came to the North American continent, they considered it a wilderness, something wild, savage, and untamed that demanded the imprint of "civilized" activity to bring it into conformance with a world made for people. Alexis de Tocqueville recognized this dominant perspective while traveling in the new nation in 1831, observing that Americans were not interested in the beauty of the wild forests and rivers, but rather that "their eyes are fixed upon another sight, the march across these wilds, draining swamps, turning the course of rivers, peopling solitudes, and subduing nature" (de Tocqueville, 1953, p. 47). In 1903 Mark Twain quoted someone else who in 1837 stated that the Mississippi River was "a river of desolation, and instead of reminding you, like others, of an angel which has descended for the benefit of man, you imagine it a devil" (Twain, 1903, p. 295).

Throughout the early history of the country, this perspective of society overcoming the resistance of nature and turning natural resources to human good was virtually the only public policy ethic for rivers. Through the very earliest times, rivers powered gristmills and embryonic industries, and diversions and water control structures often supplied agriculture during the colonial period. During the first half of the 1800s, the construction of canals added artificial components to the natural channel systems, and from midcentury onward, the alteration of rivers to obtain water for mining and lumbering was common. By the end of the 1800s, the installation of dams for major irrigation projects on the Plains and in the far West became major features of environmental, economic, and social landscapes. It was not until the dawn of the twentieth century that the conservation of resources, including water, reached the national agenda as the growing population and economic development associated with it put increased strains on natural resources (Huth, 1957).

The emphasis in the early twentieth century was still on use of environmental resources to meet human-centered needs; the question was not so much preservation versus development as it was how to develop the resources. The emphasis was on wise, forward-looking use, with the conservation of resources such as water and rivers viewed as sound business practice (Hays, 1959). This civic-minded approach to resource development shared the public stage with engineering, which enjoyed great recognition and esteem. Engineering accomplishments in transportation, such as railroads, and in industries ranging from steel making to textiles, brought a new prominence to the engineering profession (Barry, 1997). Public confidence in engineering extended to hydrologic and civil engineers who were designing and building river works. Engineers became national heroes. John S. Eastwood, for example, was viewed as a savior in the West, where he perfected the design and construction of multiple concrete arch dams that served rural and urban interests alike (Jackson, 1995). Herbert Hoover, the nation's leading engineer, became president.

This emphasis on taming lands and rivers as part of America's destiny continued largely unabated until the late 1960s, when national policymakers became more concerned with environmental quality (Harper, 2001). Previously, the installation of dams had been resisted by some local populations, largely because of the drowning of agricultural lands by reservoirs or loss of personal property, as on rivers of the Tennessee system when the Tennessee Valley Authority began its dam construction program in the 1930s (Cutler, 1985). In the 1960s, however, rivers were pivotal in the adjustment of American social values, especially those related to water quality. Urbanization, industrialization, and overuse had degraded water quality by the 1960s to a degree noticeable by ordinary citizens, and widely publicized events such as the Cuyahoga River catching fire in Cleveland, Ohio, began to highlight the costs of using water and rivers as resources without regard for the consequences. Although there was little understanding of the downstream outcomes of dams at the time, the 1960s and 1970s saw the emergence of national policies to restrict dam construction (e.g., the Wild and Scenic Rivers Act of 1968) and to improve water quality in rivers (e.g., the Clean Water Act of 1972 and later amendments).

The 1970s and later decades witnessed the emergence of a broadly defined set of social values in the United States that placed emphasis on the preservation and restoration of environmental quality in general (Harper, 2001), and the quality of river environments in particular. The desire for aesthetically pleasing environments along streams, the use of rivers in historical restoration efforts, and the central position of rivers in the maintenance of endangered species all fit into a perspective that sought an improved balance between economic development and environmental quality. Today, however, amid increasingly diverse opinions, consensus social values for rivers and their dams are difficult to define; the nearly monolithic opinions of a century ago are gone. As a result, a decision about whether or not to remove a dam depends not only on the forecasted outcomes for the physical, biological, and economic systems, but also on public perceptions of the future aesthetics and use of the resulting landscape.

In one of the few windows on public opinion about dam removal, Born et al. (1998) conducted surveys showing the trade-offs people make when a dam is removed (Box 7.1). Investigations of 14 cases of dam removal show that the public perceived important losses when dams were removed, including the loss of hydropower and recreational activities associated with reservoirs. The public also perceived important gains such as improved safety and savings related to discontinued maintenance. The survey showed not only that there was a variety of opinions, but also that in half the cases the public perceived an important gain to improved fish and wildlife habitat offered by a free-flowing river. The valuation by public opinion of a free-flowing river is relatively new on the river manage-

Box 7.1 Perceived Gains and Losses Resulting from the Dam Removals in Wisconsin

Research in Wisconsin shows the range of social and cultural values associated with recent dam removals. Born et al. (1998) summarized the perceived gains and losses associated with 14 dam removals in the state as follows. For the majority of cases, members of the public mentioned as a social value the loss of recreational activities and aesthetics associated with the impoundment. Over half mentioned the loss of a nostalgic location and loss of fish and wildlife habitat associated with the impoundment. Other concerns were the loss of potential hydropower generation, even if the dam had not generated electricity for decades and major retrofits would have been necessary. Reduced property value of lakefront development was noted in four cases as a significant issue in dam removal. Stakeholders in 12 of the 14 cases listed safety improvements as a gain, and half listed improved fish and wildlife habitat as a result of a free-flowing river. The elimination of maintenance and potential liability for owners was seen as a positive result of removal at many sites. Improved recreational activities associated with a free-flowing river were valued in only three of the cases; three also valued the improved aesthetics of a free-flowing river.

Box 7.1 continued

Perceived Gains and Losses	Dam Removals: Fulton	Greenwood	Hayman Falls	Huntington	Lemonweir	Manitowoc Rapids	McClure	Nelsonville	Ontario	Prairie Dells	Pulcifer	Somerset	Woolen Mills	Young America
Perceived losses														
Recreational activities associated with the impoundment	x	x		x	x	x	x	x	x	x	x	x	x	x
Aesthetics of the impoundment	x		x	x		x	x	x	x	x		x	x	x
Fish and/or wildlife habitat associated with the impoundment	x	x		x	x	x	x		x	x	x			x
Nostalgic location	x			x	x		x	x	x	x			x	
Potential to generate "clean" hydroelectric power	x		x	x			x			x	x			
Historical structure, point of interest		x	x			x		x	x	x				
Lakefront property values						x		x		x				
Perceived gains														
Safety improvements	x	x	x	x	x	x	x	x	x	x	x	x		
Improved fish and wildlife habitat of a free-flowing river					x	x		x	x	x		x	x	x
Perpetual maintenance costs eliminated	x		x	x	x		x	x		x				
Liability risks eliminated	x		x	x	x	x	x			x				
Improved recreational activities assocaited with a free-flowing river									x			x	x	
Improved aesthetics of a free-flowing river						x			x	x				
Fulfilled legal obligation of administrative order									x					x
Prevention of shoreline erosion		x												

Source: Adapted from Born et al., 1998.

ment and decision making landscape, and it represents a very different perspective than emphasis on water resource development that prevailed unchallenged just a century ago. Nonetheless, the results of the surveys with their wide range of opinions and perspectives show why dam removal decisions are rarely straightforward and unambiguous.

There are few reliable measures for social values, but public opinion about maintaining or removing a dam can be ascertained through surveys. One potential route to assessing public opinions is a method used by the late Professor Marie Morisawa, a geologist at the State University of New York at Binghamton. She provided respondents with pictures of various river landscapes and asked them to rate the scenes according to their desirability as recreation sites. A similar approach can be informative in making dam removal decisions. In considering the removal of a dam, an inevitable question arises among local residents about what the river landscape will look like after the dam is removed. The digital alteration of photographs of the exact sites, or photographs of similar sites without dams, can provide comparisons with existing conditions with the dam in place. Standard survey techniques, such as the assurance of a broadly representative sample of the affected population, can ensure a reading of existing social values in the locale of the potential dam removal.

DAMS AND TRIBAL CULTURE IN THE UNITED STATES

Tribal culture and religion are attentive to the human place within nature. Perhaps nowhere are the differences in tribal and nontribal ways of relating to nature more evident than in the treatment of water. Many traditional Native Americans had, and continue to have, a perspective on nature that is very different from the Jewish and Christian ethic common in the early decades of the nation's history. Many Native American religions view nature not as something to be overcome, but as something to be blended with and respected. In this view, people are part of a matrix or society of natural objects and places that puts humans on an equal footing with nature rather than in a superior position (Callicott, 1982). The emphasis on the importance of natural objects and places does not mean that tribes have been averse to changing or manipulating their environment, because when and where possible they effected significant changes (Martin, 1981). As traditional values enjoy resurgent positions in many tribes, there is a renewed awareness of the importance of nature, and new efforts are under way to reduce damaging impacts of human activities.

Dams play an important role in affecting rivers and the natural objects associated with them from a Native American and perspective. For example, coastal tribes in the Pacific Northwest for economic as well as cultural reasons revere many fish species, such as migrating salmon and sea lamprey. Many tribes in the region once subsisted on salmon as a primary food source, so the installation of dams that disrupted the migration had serious dietary consequences.

Because of the value of nature in traditional Native American culture, there are certain places in the geographical landscape that have special importance. These areas are sacred spaces, revered for their particular connection to the spiritual world. Native Americans are not alone in this definition of sacred spaces, because many cultures set aside special areas or sites for religious or spiritual reasons (Graber, 1976). Riverine environments often play a special role for Native Americans, just as was the case elsewhere in the ancient world (Tuan, 1974). Dams affect such sacred spaces by drowning them under reservoir waters, or by otherwise altering them through changes in downstream flows.

Historically, small dams were likely to be constructed on significant Native American sites. Small modern structures have the same purposes as did dams built prior to European settlement: to divert flow from sites where the river morphology constricted flow, or to raise water levels at sites that previously were logical choices for fish weirs. Therefore, the sites of the dams themselves may be significant for the Native American population. At present, there are insufficient data on the locations of dam sites on Native American land to evaluate their significance; such data will need to be collected on a case-by-case basis.

Many traditional Native American religions ascribe sacred values to water features in the general landscape. For example, among the Native Americans at Taos Pueblo in New Mexico, many lakes, springs, and streams in the vicinity of the pueblo are places to which individuals must go to perform rituals at particular times of the year (Bodine, 1979). Any interference with these sites, either through dam construction or because of dam removal, is of concern to the pueblo residents.

Dams and their reservoirs may exert strong influences on specific cultural sites. Three examples are two large dams, Glen Canyon and Fort Peck dams, and the medium-sized Elwha Dam. Glen Canyon Dam created Lake Powell, which inundates hundreds of archeological sites important to the Navajo tribe and influences downstream conditions near salt mines in the Grand Canyon held sacred by a number of tribes, including the Navajo, Hopi, and Zuni. Fort Peck Dam inundates sacred lands of the Sioux tribe in the Dakotas, and Elwha Dam creates an artificial lake over the place of origin for the Elwha Indians. Numerous dams across the entire nation are entangled in one way or another with tribal sacred space (Waldman, 1985). If nothing else, any dam removal that resurrects land that previously was inundated by reservoir waters needs to take into account potential Native American land claims, an important subject to tribes and a defining political issue for them.

Defining sacred space for protection by members outside individual tribes is problematic. Native Americans are reluctant to reveal the exact location of many sacred sites, fearing the interference of curiosity seekers or the removal of temporary shrines constructed close by. The dominant Anglo-American population, although increasingly sensitive to Native Americans' point of view, does not have a track record of protecting Native American interests, especially in cases in which resource development would result in economic gain for the majority. Therefore, Native American input into decision making about dam removal may be less forthcoming or precise than some other inputs.

Sources of information about Native American concerns regarding sacred space also may be difficult to assess. Tribal representatives may simply designate large areas rather than specific sites as being of religious interest so as not to reveal too much detail about the sites. Oral histories, written historical accounts and observations, and consultation with tribal representatives are all sources of information that may be useful in alerting decision makers to potential conflicts between Anglo-American and Native American values. Native American societies are just as complex as Anglo-American society, and there are likely to be differences of opinion within tribal communities about the outcomes of dam removal, particularly if the benefits are perceived as largely related to religious concepts rather than to economic gain for tribal members.

SOCIAL IMPACT ASSESSMENTS OF DAM REMOVAL PROJECTS

Social impact assessments in the United States developed out of a need to apply the knowledge of sociology and other social sciences to predict the social outcomes of development projects subject to the National Environmental Policy Act (NEPA) legislation of 1969 (Burdge 1994). A goal of social impact assessments (SIAs) is to identify and understand the consequences of change for human populations and communities. Unfortunately, the use of social impact assessments has not been consistent (Burdge 1994). The objective in the SIA process is to anticipate and predict social impacts in advance, so that findings and recommendations can become part of the planning and decision-making process.

Burdge (1994) developed a list of 26 social variables that may be useful in conducting a social impact assessment (Box 7.2). These variables represent the types of outcomes arising from planned change in local communities. Burdge used the following criteria in the selection of the 26 SIA variables:

1. An SIA variable is operative when a community may be altered by project development and policy change.
2. An SIA variable will tell the decision maker or planner a specific consequence of the proposed action.
3. An SIA variable always has a discrete, nominal, or continuous empirical indicator that can be measured, collected, and interpreted within the context of a specific social impact setting.
4. All SIA variables are based upon data that can be collected or made available during the planning and decision stage as well as other stages in the development of the project or policy.
5. An SIA variable does not require, but may utilize, information from questionnaires of the general population.
6. An SIA variable is not to be confused with sociological labels such as middle class, ethnicity, or other small groups.

To provide an illustration of how the SIA variables might help to inform a decision about whether or not to remove a dam, the Heinz Center panel adapted a table from Burdge (1994) (Table 7.1).

Sometimes the social process of reaching a decision to remove a dam can be a catalyst for bringing diverse interests together rather than sharpening their differences. Citizen's groups such as watershed councils often are formed to address a single issue (such as dam removal) and later evolve to address wider-scale issues (National Research Council, 1999). An example is the broadly defined effort by local citizens to organize and raise money for the removal of Stoever's Dam in Pennsylvania (Box 7.3).

Box 7.2 Social Impact Assessment Variables

Population Impacts

- Population change
- Influx or outflux of temporary workers
- Presence of seasonal (leisure) residents
- Relocation of individuals and families
- Dissimilarity in age, gender, racial, or ethnic composition

Community/Institutional Arrangements

- Formation of attitudes toward the project
- Interest group activity
- Alteration in size and structure of local government
- Presence of planning and zoning activity
- Industrial diversification
- Enhanced economic inequities
- Change in employment equity of minority groups
- Change in occupational opportunities

Conflicts between Local Residents and Newcomers

- Presence of an outside agency
- Introduction of new social classes
- Change in the commercial/industrial focus of the community
- Presence of weekend residents (recreational)

Individual and Family-Level Impacts

- Disruption in daily living movement patterns
- Dissimilarity in religious practices
- Alteration in family structure
- Disruption in social networks
- Perceptions of public health and safety
- Change in leisure opportunities

Community Infrastructure Needs

- Change in community infrastructure
- Land acquisition and disposal
- Effects on known cultural, historical, sacred, and archaeological resources

Source: Burdge (1994) p. 37.

In this instance, the dam was rebuilt and its reservoir made into a public park. A similarly satisfactory, if opposite, outcome resulted in the town of West Bend, Wisconsin, where the Woolen Mills Dam was removed through the cooperative efforts of the state Department of Natural Resources, the dam owner, local citizens, and businesses. The reservoir

Table 7.1 Social Impact Assessment Variables by Dam Removal Project Stage

Project Stage	Variables
Problem Identification	Small or medium-sized dams
Planning/Policy	Safety issues, environmental issues, public attitudes toward the project
Implementation	Relocation of families, influx of workers, change in recreation, change in wildlife habitat
Maintenance	Safety problems, insurance liabilities, flood protection, fish passage issues, native versus non-native values, sediment removal
Removal	Changes in employment, potential change in property value, restoration of natural environment, hydropower replacement, native fish return, introduced fish species leave, flood protection needed for houses near river, sediment flows to beaches, sediment quality and removal issues

Adapted from Burdge, 1994.

became the site of an extensive public park and recreation facility that includes a reconstructed trout stream.

CONCLUSION AND RECOMMENDATIONS

There is little research on social science aspects related to dam removal. This is a serious shortcoming because the social context of dam removal decisions is often as important as the environmental and economic contexts. The lack of relevant social science research is a serious shortcoming in the knowledge base for dam removal decisions because these decisions are made by people and affect them as much as they do the environment. This significant gap could be filled in many ways. For instance, research in sociology, geography, history, and planning could investigate the connections among communities, rivers, and dams. There is also more to learn more about the cultural significance of dams. Some dams or structures directly associated with them may have substantial historical signifi-

Box 7.3 Saving Stoever's Dam in Pennsylvania

The case of Stoever's Dam illustrates how a dam removal decision-making process can bring a community together. In the 1920s, the Union Canal Company needed a steady supply of water in its canal and built Stoever's Dam across a tributary to Brandywine Creek in Lebanon County. More than 150 years later, in 1978, the dam was identified as a high hazard facility after an investigation required by The Dam Inspection Act (P.L. 92-367) discovered that the dam' storage and spillway capacity was capable of passing only 11 percent of the probable maximum flood load. The investigation also found the dam and its appurtenant structures in poor condition and concluded that a failure would affect several inhabited structures located downstream.

Stoever's Dam in 1915.

Courtesy of Pennsylvania Department of Environmental Resources

In 1981, the Pennsylvania Department of Environmental Resources, Division of Dam Safety, requested that the dam's current owner (the city of Lebanon) either drain the reservoir or present a proposal for the dam's rehabilitation. Although the local government and citizens favored saving the dam and creating a focal point around the reservoir for a new park, the city did not budget the money necessary to rebuild the dam. The city, along with local citizens, organized the Save Stoever's Dam Committee to garner political and monetary support to rebuild the dam. In addition to applying for state and federal money, the committee sponsored various fundraisers. Several local sportsmen and youth groups also held fundraisers.

The rebuilding project began in April 1982 and was completed in January 1983. The final project costs, including design and construction, totaled more than $605,000. Funding was provided by $500,000 in grants from the Pennsylvania Department of Community Affairs and the U.S. Department of Housing and Urban Development. The Save Stoever's Dam Committee raised the remaining money, more than $100,000, needed to complete the project.

cance, so there may be reasons to remove only part of a dam or to preserve or restore the associated mill works or power house. A particularly important line of investigation that could be undertaken by nongovernmental organizations with the cooperation of state agencies would be to investigate the social and economics outcomes after dam removal. These after-project studies are at least as important as environmental and social impact studies undertaken before the dam removal.

Although the social outcomes of dam removal decisions are not yet well known, standard survey-based research outlines changes in individual and community behavior related to such decisions. Adaptive management for environmental systems could be extended to social systems, so that river managers could make informed adjustments to their plans. In this way, an active role would be reserved in the decision-making process for people and communities as well as natural organisms and ecosystems.

- **Conclusion:** Social science perspectives on dam removal suffer from a lack of research on the subject, so that decision makers wishing to include social perspectives in the process are faced with many unknowns and little guidance.
- **Recommendation:** The panel recommends that agencies and organizations that fund social science research support investigations into the social and cultural dimensions cases in which dams have been removed as a way of improving the predictability of outcomes.
- **Recommendation:** The panel also recommends that decision makers in dam removal cases should undertake social impact studies modeled on the environmental impact studies that are a common feature of such decision-making processes. These social impact studies should address the cultural significance of the dam site (e.g., as a tribal sacred site), reservoir area, and river areas likely to be changed by the proposed removal.

The Future

The familiar wisdom is that times change, and that time waits for no one. During the deliberations that led to this report, a number of changes occurred on the national science and policy landscape that are relevant to the future of dam removal. Severe electrical power shortages in California and elsewhere prompted new interest in power sources, including dams. Terrorist attacks on the nation's civilian populations and concern for the water control infrastructure, especially dams, became front-page news. A new administration assumed the presidency and the executive branch of the government, bringing with it changes in perspective and opinion. Control of the U.S. Senate switched from one political party to another. At the same time, new advances occurred on the scientific front. Biologists discovered fishes (the robust redhorse) thought to be extinct but apparently surviving in southeastern rivers with dams. Physical and biological scientists were cooperating in research focusing on the effects of dam removal in locales as diverse as Pennsylvania and California, and their reports were beginning to work their way through to publication in the refereed journal literature. New knowledge was beginning to surface to deal with old problems. Thus, managers, scientists, and decision makers were gaining a larger experience base for dam removals.

This report is a primer for researchers, decision makers, and the public, providing information on background, science, and decision concepts. If funding is provided for a second phase of the project, the Heinz Center will conduct a scientific conference on research related to the outcomes of dam removal. The conference will feature invited speakers from the major scientific disciplines engaged in research related to dam removal and provide them with the opportunity to interact with each other in accomplishing three tasks: (1) outlining the present state of knowledge in their individual fields, (2) identifying those topics in one field that are in need of supporting research from other fields, and (3) specifying gaps in scientific knowledge that require additional research to support decision making. The proceedings of the conference will include written papers and conclusions. In addition, if funding is provided, the Heinz Center may experiment with the decision-making processes outlined in this report in two or more localities dealing with dam removal.

The present report is a summary of current understanding of the decision-making process regarding dam removals that are scientifically informed and serve the best interests of the largest number of stakeholders. It represents the state of the knowledge about these decisions at one point in time. The possible futures for the nation's rivers turn on the wise use of knowledge about them, and the wise use of them as resources. Because experience is an important part of that wisdom, subsequent advances will augment and change some of the concepts in this report. The greatest chance for success in making good decisions about dam removal lies in the use of this report as merely a starting point, and in the subsequent growth and change of the ideas it presents.

Appendixes

Appendix A

USEFUL SOURCES OF INFORMATION ON THE WORLD WIDE WEB FOR DAM REMOVAL DECISION MAKERS

EXISTING DAMS

National Inventory of Dams (U.S. Army Corps of Engineers and Federal Emergency Management Agency)

http://crunch.tec.army.mil/nid/webpages/nid.cfm. This web site (as of Oct. 15, 2001) is offline as a security precaution in light of the Sept. 11, 2001, terrorist attacks on the United States. The site may be restored after further evaluation.

- The online interactive map and downloadable database contains information about approximately 76,000 dams. Includes those structures whose collapse might be a threat to life and property downstream, those greater than 6 ft (2 m) high with more than 50 acre-ft (61,000 cu m) of storage, and those that are 25 ft (8 m) high with more than 15 ac ft (18,500 cu m) of storage.

National Atlas (U.S. Geological Survey)

http://www.nationalatlas.gov/damsm.html

- This online interactive map and downloadable database contains information about approximately 7,700 major dams. A subset of the National Inventory of Dams, the dataset includes dams that are 50 feet or more in height, have a normal storage capacity of 5,000 acre-feet or more, or have a maximum storage capacity of 25,000 acre-feet or more.

DAM REMOVALS

Dam Removal Success Stories (American Rivers, Friends of the Earth, and Trout Unlimited)

http://www.americanrivers.org/damremovaltoolkit/successstoriesreport.htm

- In December 1999, these three organizations issued a cooperative report outlining the experiences of specific dam removal projects. American Rivers also has a resource center of material regarding dam removals at http://damremoval.americanrivers.org.

Wisconsin Rivers (Wisconsin River Alliance)

http://www.wisconsinrivers.org/

- This organization provides examples of changes brought about by dam removal and useful information on dam removal decision-making processes.

MAPS

National Atlas (U.S. Geological Survey)

http://nationalatlas.gov

- The atlas contains a variety of high-quality, small-scale maps for the entire United States, including authoritative national geospatial and geostatistical data sets. Examples of digital geospatial data include soils, county boundaries, volcanoes, rivers, streams, and watersheds.

Topographic maps

U.S. Geological Survey

http://www.usgs.gov

- Paper maps can be ordered online.

MapTech

http://www.maptech.com/mapserver.

Microsoft's Terraserver

http://terraserver.microsoft.com

- This web site includes aerial photography for many parts of the nation.

TopoZone

http://www.topozone.com

- This web site provides digital topographic maps at a variety of scales.

Maps showing census data related to social and economic data

Bureau of the Census

http://tiger.census.gov and http://factfinder.census.gov/servlet/BasicFactsServlet

Maps showing environmental data

Environmental Protection Agency

http://maps.epa.gov/enviromapper
http://www.epa.gov/surf/

Department of Housing and Urban Development, Healthy Communities
http://www.hud.gov/emaps

U.S. Geological Survey and U.S. Environmental Protection Agency, National Hydrography Dataset
http://nhd.usgs.gov/

Maps showing earth science data
U.S. Geological Survey
http://geode.usgs.gov

Base maps for use with geographical information systems
Environmental Systems Research Institute, ArcView
http://www.esri.com/software/arcview/index.html

Map Info Professional
http://dynamo.mapinfo.com/products/web/Overview.cfm?productid=44

Hydrological Information and Maps
Federal Emergency Management Agency
http://www.fema.gov/maps

- This web site contains surveys and highly detailed topographic maps (including cross sections) of many streams and rivers used to determine the extent of the active channel and the 100-year floodplain.

National Hydrography Dataset (U.S. Geological Survey and U.S. Environmental Protection Agency)
http://nhd.usgs.gov/

- The dataset is a basic source for stream and river geography.

U.S. Geological Survey
http://www.usgs.gov/water

- A variety of water data based on the 6,600 stream gages USGS operates. Data are available for each day of record, as well as in an abbreviated form showing only annual peak flows; information on each gaging station includes its dates of operation and a map showing its precise location. Users can retrieve the data either in tabular form for numerical analysis, or in easily read graphs.

OTHER DATA SOURCES

Sediment

U.S. Geological Survey

http://water.usgs.gov/owq.html

- The USGS keeps data on the quantity of sediment discharged passing through approximately 1,600 gaging stations.

Water Quality

National Water Quality Assessment Program (U.S. Geological Survey)

http://water.usgs.gov/owq/data.html

- Near real-time and historical data for many of the nation's rivers are available from the web site. The data can be downloaded in the form of tables for analysis from much, but not all, of the country.

Appendix B
Glossary

Acre-foot the amount of water required to cover one acre to a depth of one foot. An acre-foot equals 326,851 gallons or 43,560 cubic feet.

Adaptive management a systematic process for continually improving management policies and practices by learning from the outcomes of operational programs. Its most effective form — "active" adaptive management — uses management programs that are designed to experimentally compare selected policies or practices, by evaluating alternative hypotheses about the system being managed.

Aggradation the raising of a riverbed due to sediment deposit

Allochthonous characteristic of or referring to events originating from outside the organism or the self.

Anadromous fish that hatch in freshwater, migrate to the ocean to mature, and return to freshwater to spawn.

Anchoring The practice, by some survey respondents, of basing stated willingness to pay for a non-market good on the known value of a market good that is considered similar or related in some way.

Arch dam a dam construction type used at sites that are narrowly constricted (e.g., valley or canyon containing the stream) and that spans the valley opening as one single structure, anchored in the sidewalls by thrust blocks.

Autotrophic needing only carbon dioxide or carbonates as a source of carbon and a simple inorganic nitrogen compound for metabolic synthesis

Bequest value a willingness to pay to preserve the environment for the benefit of one's descendants.

Breach a break or opening in a dam

Buttress dam a dam construction type made of flat decking sloping from crest to the base, typically comprised of reinforced masonry or stonework built against concrete.

Channelization the modification of a natural river channel, including deepening, widening, or straightening

Crib dam a dam construction type that is constructed of a timber outer box typically filled with rocks for stability, sometimes further stabilized with wire or brush blankets

Dam any barrier that impounds or diverts water

Decommissioning is a term used mostly for dams that are or have been generating hydropower and are shutting down power operations after losing relicensing from the Federal Energy Regulatory Commission (FERC). This may or may not include removing diversions for power generation, shutting down operations entirely, safe maintenance of dams after turbines are shut down and restoring sites to their normal, pre-project conditions.

Dam removal removal of the entire structure on a river or stream. This can also include restoration of the site to pre-project conditions.

Diversion the taking of water from a body of water into a pipe or other conduit

Earth fill dam a dam construction type that is constructed from local earth materials that are shaped and rolled into a sill across the watercourse to be dammed.

Existence value In the case of a unique and essentially irreplaceable resource, the value experienced by some due to the simple knowledge that the resource exists, irrespective of any current or expected future use.

Erosion wearing away of the land surface by detachment and movement of soil and rock fragments during a flood or storm or over a period of years through the action of wind, water, or other geologic process.

Fish ladder a series of ascending pools of running water constructed to enable fish to swim upstream around or over a dam.

Fish passage any feature of a dam that allows fish to move around, through, or over a dam without harm.

Free-flowing Undammed and unchannelized watercourses, as defined by the federal Wild and Scenic Rivers Act.

Gravity dam a dam construction type usually made of concrete, the weight of which is capable of providing the major resistance to the water forces exerted on it.

Hydroelectric power electric power generated by a flow of water.

Hypothetical bias Random error in survey results which is attributable to the hypothetical nature of the valuation task. Hypothetical bias is not actually a bias, since it is defined to have a zero mean.

Implied value cues Information communicated, explicitly or implicitly, in the course of an interview or in the body of a survey instrument that serves to suggest a value or range of values that may be appropriate for the non-market good in question.

Impoundment a body of water (such as a pond or reservoir) confined by a dam, dike, floodgate or other barrier used to collect and store water for future use.

Levee a raised embankment of a river, showing a gentle slope away from the channel, usually built to protect land from flooding.

Nonresponse bias A systematic error in valuation which results from incorrect assumptions about respondents who do not answer some or all questions; for example, assuming that these respondents hold values similar to those who do answer the questions.

Option value often categorized as a nonuse or passive use value and refers to the fact that an individual places a certain current value on the option to use a resource in the future.

Reservoir a large natural or artificial lake used as a source of water supply.

Restoration return of an ecosystem to a close approximation of its condition before a disturbance. The goal is to emulate a natural, functioning, self-regulating system that is integrated with the ecological landscape in which it occurs.

Riparian pertaining to a river.

Riparian habitat the habitat found on the bank of a natural watercourse (as a river) or sometimes a lake or tidewater.

River a natural stream of water of considerable volume.

Rock fill dam a dam construction type that uses rocks for weight and stability with a cover or membrane to provide watertightness.

Run-of-the-river dam A structure built by humans across a river or stream for impounding water, such that the impoundment at normal flow levels is completely within the banks and all flow passes directly over the entire dam structure within banks, excluding abutments, to a natural channel downstream. Some dams with storage reservoirs create a run-of-river condition through operating rules, whereby the dam releases water at approximately the same rate as the reservoir receives it.

Sample bias A systematic error in valuation which results from the way in which the sample of respondents was selected from the population.

Spillway a channel on a dam over which excess or flood flows are discharged designed to prevent impounded water from escaping over the top of the dam.

Strategic bias A systematic error in valuation which results from attempts by respondents to answer questions in a way that will benefit them in the future; for example, by understating values so as to cause lower user fees.

Stream Order The numbering of streams in a network. There are many different methods; the most widely used is a classification which labels all unbranched streams as first-order streams. When two first order streams meet, the resulting channel is a second-order stream. Where two second-order streams meet, a third-order stream results, and so on. Any tributary of an order lower than the main channel is ignored.

Watershed a region or area bounded peripherally by a divide and draining ultimately to a particular watercourse or body of water.

Appendix C

ABOUT THE CONTRIBUTORS AND THE PROJECT STAFF

The Heinz Center panel met with federal and state officials during a field visit to the site of Rindge Dam on Malibu Creek in California. Back row, left to right: Syd Brown (California Department of Parks and Recreation), David Wegner (panel member), David Freyberg (Stanford University), Reinard Knur, Sheila David (Heinz Center Fellow and project manager), Will Graf (panel chair), Tom Downs, Mary Lou Soscia (panel member), William J. Bennett (California Department of Water Resources), Jack Kraeuter (panel member), Phil Williams (panel member), Doug Dixon (panel member). Front row: Jason Shea (U.S. Army Corps of Engineers), Chris Peregrin (California Department of Parks and Recreation), Suzanne Goode (California Department of Parks and Recreation), and Robert Hamilton (U.S. Bureau of Reclamation). Photograph courtesy of Sarah Baish.

THE CONTRIBUTORS

William L. Graf, *Chair*, is Educational Foundation Endowed University Professor and professor of geography at the University of South Carolina. His specialties include fluvial geomorphology and policy for public land and water. His research and teaching have focused on river-channel change, human impacts on river processes, morphology, and ecology, as well as contaminant transport and storage in river sediments. In the arena of public policy, he has emphasized the interaction of science and decision making, and the resolution of conflicts among economic development, historical preservation, and environmental restoration for rivers. He has published several books and more than 100 scientific papers and is past president of the Association of American Geographers. Previously, he was Regents Professor of Geography at Arizona State University. He has served on the National Research Council (NRC) as a member of the Water Science and Technology Board and is a member of the Board on Earth Sciences and Resources. His NRC committee experience includes a 10-year membership on the Committee to Review the Glen Canyon Environmental Studies and chairing the Committee on Innovative Watershed Management and the Committee on Research Priorities in Geography at the U.S. Geographical Survey. He was appointed by President Clinton to the Presidential Commission on American Heritage Rivers. He is also a National Associate of the National Academy of Science.

John J. Boland is a professor of geography and environmental engineering at Johns Hopkins University. He is a registered professional engineer. His background includes management positions in water and wastewater utilities, teaching, research, and consulting activities at all levels of government and private industry. Dr. Boland has published widely on economic aspects of water and resource policy. He is an associate editor of *The Annals of Regional Science* and a member of the Risk Management Technical Advisory Workgroup of the American Water Works Association. He has served on a number of committees and panels of the National Research Council and served as chair of the Water Science and Technology Board (1985–1988). He holds a B.E.E. in electrical engineering from Gannon College, an M.S. in governmental administration from George Washington University, and a Ph.D. in environmental economics from Johns Hopkins University.

Douglas A. Dixon manages the aquatic protection, water quality and fishery research initiatives at the Electric Power Research Institute (EPRI). Dr. Dixon has more than 25 years of wide-ranging experience in environmental science and engineering research, including 10 years assessing the impacts of hydroelectric projects on aquatic resources. He is experienced in the regulatory and procedural requirements of the Clean Water Act, National Environmental Policy Act, Federal Power Act, Endangered Species Act (ESA), and numerous other environmental acts as they apply to the Federal Energy Regulatory Commission's licensing and re-licensing authority and state permitting of steam-electric facilities. Areas of expertise include ecological risk assessment, environmental impact analysis, ecological modeling, fish passage, impingement/entrainment monitoring, ESA assessments, instream flow assessment, fisheries management plan review, and current and historical aquatic resource assessment. He holds a B.A. in biology from the State University of New York at Geneseo and a Ph.D. in marine fisheries science from the College of William & Mary.

THOMAS C. DOWNS is a member of Patton Boggs's administrative and regulatory practice and concentrates on environmental law and general public policy. Mr. Downs has extensive experience in federal environmental law and policy, including the Comprehensive Environmental Response, Cleanup, and Liability Act (CERCLA, or Superfund); brownfields; Clean Water Act; and solid waste issues. Mr. Downs assists clients in briefing congressional offices, drafting legislation and congressional briefing papers, and developing legislative strategy. He also works closely with federal agency staff on executive branch initiatives. Mr. Downs handled environmental and natural resources legislation and other matters on Capitol Hill for more than 10 years before joining Patton Boggs. He served as legislative director and chief of staff to Representative George J. Hochbrueckner (NY-01) from 1987 to 1995 and as legislative assistant to Representative Martin O. Sabo (MN-05) from 1985 to 1987. He received a B.A. from Brown University in 1983 and J.D. from The American University, Washington College of Law in 1994.

JOHN J. KRAEUTER is an aquatic biologist with the Pennsylvania Department of Environmental Protection, Bureau of Waterways Engineering. His responsibilities include evaluating the environmental impacts of the proposed construction, modification, and removal of dams across the Commonwealth. Previously, he worked for a consulting firm specializing in the collection of biological data, including fishery, benthic macroinvertebrate, plankton, and freshwater mussel surveys, in streams, rivers, and lakes. He holds a bachelor's degree in biology from the University of Delaware.

MARY LOU SOSCIA serves as the Columbia River coordinator for the U.S. Environmental Protection Agency (EPA), Region 10-Seattle. In this role, she represents the EPA in discussions of the role of the Clean Water Act in future federal decisions on the Columbia River power system. Ms. Soscia has more than 20 years of experience with state, federal, and tribal government, specializing in watershed and river management issues. While on an assignment from the EPA in 1993–1997, Ms. Soscia served as coordinator of the Tribal Watershed Program for the Columbia River Inter-Tribal Fish Commission and as program manager for the Oregon Watershed Health Program. While working for the EPA in Washington, D.C., Ms. Soscia served on the team that developed and established the National Estuary Program, a collaborative effort to restore the nation's estuaries. Ms. Soscia also has worked for the states of Maryland and Wyoming. Ms. Soscia has a bachelor's of science degree in geography from Virginia Polytechnic Institute and State University and a M.S. in geography from the University of Maryland.

DAVID L. WEGNER has been involved in the design, coordination, and implementation of innovative scientific and river rehabilitation programs in the western United States and internationally since the late 1970s. Recently he established a company, Ecosystem Management International, Inc., that applies the scientific expertise developed in the Grand Canyon to river and terrestrial rehabilitation work focusing on endangered species and river process studies. From 1982 through 1996, Mr. Wegner coordinated the most extensive series of ecosystem studies and rehabilitation work ever attempted in the Colorado River basin. His expertise is in the areas of aquatic ecology, river engineering, and the application of science to risk assessment and adaptive management. His professional career includes work with the states of Minnesota and Utah and the U.S. Department of the Interior (DOI), and consulting with numerous Native American and environmental groups. He has received numerous commendations for public service, including recognition from the National Research Council, and is a recipient of the DOI's Bureau of Reclamation's Resource Management Award.

PHILIP B. WILLIAMS is president of Philip Williams & Associates LTD, an engineering and environmental planning firm he formed in 1979 that has offices in California, Oregon, and Washington. A professional civil engineer, he founded the San Francisco–based International Rivers Network in 1985. Dr. Williams has been engaged in a wide range of national and international hydrologic and engineering hydraulics work, primarily assessing the environmental effects of hydrologic change caused by dams and diversions and preparing feasibility studies, management plans, and environmental impact studies, related to river and wetlands restoration. He has directed more than 250 such studies, including projects on flood control, wetland restoration, river management, national park plans, water resources development, and estuarine management plans. He has authored or co-authored numerous papers on river management. He holds a Ph.D. in Hydraulics from the University of London's University College Civil and Municipal Engineering Department.

CRAIG S. WINGO is the national director for both the earthquake and dam safety programs within the Federal Emergency Management Agency (FEMA). He is responsible for the coordination of earthquake and dam safety activities at the federal level in partnership with the states and private sector. Previously, Mr. Wingo served as the deputy associate director of the Mitigation Directorate (1996 to 1999), where he had oversight responsibilities for special projects covering a broad spectrum of FEMA's mitigation programs. Prior to that, he directed the Infrastructure Support Division in FEMA's Response and Recovery Directorate, served as the assistant associate director of the Office of Technological Hazards, and held several positions in the National Flood Insurance Program. Prior to his federal service, Mr. Wingo worked in a private civil engineering firm in Maryland. He is a licensed professional engineer in the Commonwealth of Virginia and received numerous awards and citations during his federal career, including the Senior Executive Service's Meritorious Executive Award and two FEMA Meritorious Service Awards.

EUGENE P. ZEIZEL has more than 22 years of service with the Federal Emergency Management Agency (FEMA) and Federal Insurance Administration (FIA). Initially, he worked for more than three years as a project officer for flood insurance studies in Region VI. From 1981 through 1997, Dr. Zeizel managed FEMA's hurricane evacuation studies, conducted jointly with the U.S. Army Corps of Engineers and National Weather Service. He later was the project engineer for Regions VII and X, managing and resolving appeals and protests to the FIA's flood insurance rate maps. Dr. Zeizel has been working in the National Dam Safety Program Office since January 1999 and is responsible for dam safety research and training projects. He is a member of the subcommittees on research and training of the Interagency Committee on Dam Safety. Dr. Zeizel holds a B.S. in geology, M.S. in hydrology, and Ph.D. in civil engineering specializing in water resources and planning.

HEINZ CENTER STAFF

SHEILA D. DAVID is a fellow and project manager at The Heinz Center, where she is managing studies for the Sustainable Oceans, Coasts, and Waterways Program. At The Heinz Center, she has helped produce two reports: *The Hidden Costs of Coastal Hazards* and *Evaluation of Erosion Hazards.* Before joining The Heinz Center in 1997, she was a senior program officer at the National Research Council's Water Science and Technology Board for 21 years, where she was the study director for approximately 30 committees that produced reports on topics such as managing coastal erosion, restoration of aquatic ecosystems, protection of groundwater, wetlands characteristics and boundaries, water quality and water reuse, natural resource protection in the Grand Canyon, and sustainable water supplies in the Middle East. Ms. David has served as an advisor and board member of the Association for Women in Science (AWIS) and as editor of *AWIS* magazine. She is also a founder of the National Academy of Sciences' annual program honoring women in science.

Sarah K. Baish, now working towards a master's degree in urban and environmental planning at the University of Virginia, was a research associate for The Heinz Center's Sustainable Oceans, Coasts, and Waterways program through January 15, 2002. Before joining the Center, she worked in a national park in Slovakia as an environmental management consultant with the Peace Corps. Her primary responsibilities included grant writing, organizing educational events, promoting interpretive visitor services, and developing international collaborations. Before that, she had interned with the National Oceanic and Atmospheric Administration, and her work contributed to the establishment of a humpback whale sanctuary in Hawaii. Ms. Baish holds a B.A. in environmental science from the University of Virginia.

Judy Goss is a research assistant for The Heinz Center's Sustainable Oceans, Coasts, and Waterways program. She graduated cum laude with a degree in political science from Mary Washington College in May of 2001. She also works for Mary Washington as a part-time assistant debate coach. She is particularly interested in the intersection of gender and political communication, and she plans to pursue a graduate degree in communication studies.

References

Abell, R.A., D.M. Olson, E. Dinerstein, P.T. Hurley, J.T. Diggs, W. Eichbaum, S. Walters, W. Wettengel, T. Allnutt, C.J. Loucks, and P. Hedao. 2000. Freshwater Ecoregions of North America—A Conservation Assessment. Washington, DC: World Wildlife Fund and Island Press.

Allan, D.J. 1995. Stream Ecology: Structure and Function of Running Waters. London: Chapman & Hall.

American Fisheries Society (AFS). 1985. Proceedings of the Symposium on Small Hydropower and Fisheries, May 1–3, 1985 Aurora, CO. Bethesda, MD: American Fisheries Society.

American Fisheries Society (AFS). 1993. Fish Passage and Technology: Proceedings of a Symposium. Bethesda, MD: American Fisheries Society.

American Rivers. 2000. Paying for Dam Removal: A Guide to Selected Funding Sources. Washington, DC: American Rivers.

American Rivers. 2001a. Conestoga River, PA—Removal of seven dams. Obtained online at http://www.americanrivers.org/tableofcontents/ssconestoga.htm, June 28, 2001.

American Rivers. 2001b. A History of Edwards Dam Removal Efforts—The FERC Relicensing Process. Obtained online at http://www.americanrivers.org/tableofcontents/history.htm. December 13, 2001.

American Rivers, Friends of the Earth, and Trout Unlimited. 1999. Dam Removal Success Stories: Restoring Rivers Through Selective Removal of Dams that Don't Make Sense. Washington, DC: American Rivers. Also available online at http://www.americanrivers.org/damremovaltoolkit/successstoriesreport.htm.

American Sportfishing Association. 2001. The Economic Importance of Sport Fishing: Economic Data on Sport Fishing Throughout the Entire United States. Obtained online at http://www.asafishing.org/statistics/reports/economicimpact.htm, October 11, 2001.

Andrews, E.D. 1986. Downstream effects of Flaming Gorge reservoir on the Green River, Colorado and Utah. Geological Society of America Bulletin 97: 1012–1023.

Aspen Institute. 2000. Dialogue on dams and rivers description, unpublished. Aspen Institute, Washington, DC.

Auble, G.T., J.M. Friedman, and M.J. Scott. 1994. Relating riparian vegetation to present and future streamflows. Ecological Applications 4: 544–554.

Bain, M.B., J.T. Finn, and H.E. Booke. 1988. Streamflow regulations and fish community structure. Ecology 69(2): 382–392.

Barry, J.M. 1997. Rising Tide—The Great Mississippi Flood of 1927 and How It Changed America. New York: Simon and Schuster.

Bates, K. 1993. Fish passage policy and technology. Proceedings of a Symposium sponsored by the Bioengineering Section of the American Fisheries Society, September 1993, Portland, OR. Bethesda, MD: American Fisheries Society.

Baudo, R., J. Giesy, and H. Muntau (editors). 1990. Sediments: Chemistry and Toxicity of In-place Pollutants. Ann Arbor, MI: Lewis Publishers, Inc.

Baumann, W. H., H. Bayer, P. Greupner, H. Kraft, E. Lauterjung, H. Mollien, H. Wolkewitz, and G. Zeuner. 1986. Ecological Effects of Dams. Dortmund, Germany: Verkehrs- und Wirtschafts-Verlag.

Berger, J.J. 1990. Evaluating ecological protection and restoration projects: A holistic approach to the assessment of complex, multi-attribute resource management problems. Doctoral dissertation, University of California, Davis, CA.

Bodine, J. 1979. Taos Pueblo. In Handbook of North American Indians—Southwest, Vol. 9. Washington, DC: Smithsonian Institution.

Bolke, E.L., and K.M. Waddell. 1975. Chemical quality and temperature of water in Flaming Gorge Reservoir, Wyoming and Utah, and the effect of the reservoir on the Green River. U.S. Geological Survey Water Supply Paper 2039-A. Washington, DC: U.S. Government Printing Office.

Born, S.M., K.D. Genskow, T.L. Filbert, N. Hernandez-Mora, M.L. Keefer, and K.A. White. 1998. Socioeconomic and institutional dimensions of dam removals—The Wisconsin experience. Environmental Management 22(3): 359–370.

Brice, J.C. 1960. Index for description of braiding. Bulletin of the Geological Society of America 71:1833.

British Columbia Forestry Service. 2001. Definitions of adaptive management. Obtained online at http://www.for.gov.bc.ca/hfp/amhome/Amdefs.htm, October 4, 2001.

Brookes, A., and F.D. Shields Jr. 1996. River Channel Restoration—Guiding Principles for Sustainable Projects. Chichester, UK: John Wiley & Sons.

Burdge, R. 1994. A Conceptual Approach to Social Impact Assessment, Revised Edition. Middleton, WI: Social Ecology Press.

Burns, D.C. 1991. Cumulative effects of small modifications to habitat (revised). Fisheries 16(1): 12–17.

Cairns, J.J., Jr. 1990. Lack of theoretical basis for predicting rate and pathways of recovery. Environmental Management 14:517–526.

Callicott, J.B. 1982. Traditional American Indian and Western European attitudes toward nature—An overview. Environmental Ethics 4:293–248.

Camargo, J.A., and N.J. Voelz. 1998. Biotic and abiotic changes along the recovery gradient of two impounded rivers with different impoundment use. Environmental Monitoring and Assessment 50:143–158.

Capelli, M.H. 1999. Recovering endangered steelhead rainbow trout (*Oncorhynchus mykiss*) in Southern California coastal watersheds. Proceedings, Coastal Zone 99: The People, the Coast, and the Ocean. Boston: University of Massachusetts.

Carson, R.T., W.M. Hanemann, R.J. Kopp, J.A. Krosnick, R.C. Mitchell, S. Presser, P.A. Ruud, and V.K. Smith. 1996. Was the NOAA panel correct about contingent valuation? Discussion Paper 96-20, Washington, DC: Resources for the Future.

Catalano, M.J., M.A. Bozek, and T.D. Pellett. In press. Fish habitat relations and initial response of the Baraboo River fish community to dam removal. Paper presented at the North American Benthological Society meeting, LaCrosse, WI, in July 2001.

Cederholm C.J., M.D. Kunze, T. Murota and A. Sabatani. 1999. Pacific salmon carcasses: Essential contributions of nutrients and energy for aquatic and terrestrial ecosystems. Fisheries 24(10): 6–11.

Church, M. 1995. Geomorphic response to river flow regulation: Case studies and time scales. Regulated Rivers: Research and Management 11:3–22.

Clay, C.H. 1995. Design of Fisheries and other Fish Facilities. Boca Raton, FL: Lewis Publishers, 2nd Edition.

Cohen, F.S. 1982. Handbook of Federal Indian Law. Buffalo, NY: William S. Hein and Company, Inc.

Collier, M., R.H. Webb, and J.C. Schmidt. 1996. Dams and Rivers: Primer on the Downstream Effects of Dams. U.S. Geological Survey Circular 1126. Tucson, AZ: U.S. Geological Survey.

Costenbader, K. 1998. Damning dams—Bearing the cost of restoring America's rivers. George Mason Law Review 6(3): 635–673.

Coutant, C.C. (editor). 2001. Behavioral Technologies for Fish Guidance. Bethesda, MD: American Fisheries Society Symposium 26.

Cutler, P. 1985. The Public Landscape of the New Deal. New Haven, CT: Yale University Press.

Dadswell, M.J. 1996. The removal of Edwards Dam, Kennebec River, Maine: Its effects on the restoration of anadromous fishes. Draft environmental impact statement, Kennebec River, Maine, appendices 1–3, 92pp.

Dingman, S.L. 1994. Physical Hydrology. New York: Macmillan.

Doeg, T.J., and J.D. Koehn. 1994. Effects of draining and desilting a small weir on downstream fish and macroinvertebrates. Regulated Rivers: Research and Management 9:263–277.

Doyle, M.W., E.H. Stanley, M.A. Luebke, and J.M. Harbor. 2000. American Society of Civil Engineers. Joint Conference on Water Resources Engineering and Water Resources Planning and Management, Minneapolis, MN, July 30–August 2, 2000.

Dixon, D.A. 2000. Scoping Study on Sedimentation Issues at Hydroelectric Projects. Final Report TR-114008. Palo Alto, CA: Electric Power Research Institute.

Dynesius, M., and C. Nilsson. 1994. Fragmentation and flow regulation of river systems in the northern third of the world. Science 266:753–762.

Electric Power Research Institute (EPRI). 1996. Water Resource Management and Hydropower: Guidebook for Collaboration and Public Involvement. Palo Alto, CA: Report TR-104858.

Electric Power Research Institute (EPRI). 1998. Review of a Downstream Fish Passage and Protection Technology Evaluations and Effectiveness. Palo Alto, CA: Report TR 111517.

Electric Power Research Institute (EPRI). 2000a. Hydro Relicensing Forum: Relicensing Strategies. National Review Group Interim Publication. Palo Alto, CA: Report 1000737.

Electric Power Research Institute (EPRI). 2000b. Instream Flow Assessment Methods: Guidance for Evaluating Instream Flow Needs in Hydropower Licensing. Palo Alto, CA: Report 1000554.

Electric Power Research Institute (EPRI). 2000c. Scoping Study on Sedimentation Issues ay Hydropower Projects. Palo Alto, CA: Report TR 114008.

Electric Power Research Institute (EPRI). 2002. Methods for Meeting and Monitoring Dissolved Oxygen Standards for Hydroelectric Porjects. Palo Alto, CA: Report 1005194.

Environmental Defense Center. 2000. "Ventura River Listed Among Nation's Most Endangered Rivers: Matijila Dam is Obstacle to Recovery." Press Release on April 10, 2000. Obtained online at http://www.edcnet.org/press,venturariver.htm.

Environmental News Network. 1999. Oregon prepares to remove 2 dams. Obtained online at http://www.enn.com/enn-news-archive/1999/06/060299/portlanddamremoval_3454.asp, January 3, 2002.

Federal Emergency Management Agency and U.S. Army Corps of Engineers. 1994. Water Control Infrastructure—National Inventory of Dams, Updated Data, 1993–1994. http://crunch.tec.army.mil/nid/webpages/nid.cfm.

Federal Emergency Management Agency and U.S. Army Corps of Engineers. 1996. Water Control Infrastructure—National Inventory of Dams, Updated Data, 1995–1996. http://crunch.tec.army.mil/nid/webpages/nid.cfm.

Federal Energy Regulatory Commission. 1995. Preliminary Assessment of Fish Entrainment at Hydropower Projects: A Report on Studies and Protective Measures, Volumes I and II. Office of Hydropower Licensing, Washington DC. Paper No. DPR-10.

Federal Energy Regulatory Commission. 1997. Final environmental impact statement—Kennebec River Basin, Maine. Washington, DC.

Federal Energy Regulatory Commission and Electric Power Research Institute. 1996. Survey of Reservoir Sedimentation Problems in the United States. Washington, DC: Stone and Webster Engineering Corporation.

Forest Preserve District of Cook County. 1971. Nature Bulletin No. 410-A, March 13. Obtained online at http://newton.dep.anl.gov/natbltn/400-499/nb410.htm, January 4, 2002.

Freeman, A.M., III. 1993. The Measurement of Environmental and Resource Values: Theory and Methods. Washington, DC: Resources for the Future Press.

Freeman, M.C., Z.H. Bowen, K.D. Bovee, and E.R. Irwin. 2001. Flow and habitat effects on juvenile fish abundance in natural and altered flow regimes. Ecological Applications 11(1): 179–190.

Gambrell, R.P., C.N. Reddy, and R.A. Khalid. 1983. Characterization of trace and toxic materials in sediments of a lake being restored. Journal of the Water Pollution Control Federation 55:1201–1210.

Garrison, W. 1973. Disasters That Made History. New York: Abington Press.

Giesy, J., and R.A. Hoke. 1990. Freshwater sediment quality criteria: Toxicity bioassessment. Pp. 265–348 in Sediments: The Chemistry and Toxicology of In-place Pollutants. Ann Arbor, MI: Lewis Publications.

Glover, R.E. 1964. Dispersion of dissolved or suspended materials in flowing streams. Professional Paper 433-B. Washington, DC: U.S. Geological Survey.

Graber, L.H. 1976. Wilderness as sacred space. Monograph 8. Washington, DC: Association of American Geographers.

Graf, W.L. 1988. Fluvial Processes in Dryland Rivers. Berlin: Springer-Verlag.

Graf, W.L. 1993. Landscapes, Commodities, and Ecosystems: The Relationship Between Policy and Science for American Rivers. Pp. 11–42 in Sustaining Our Water Resources. Washington, DC: National Academy Press.

Graf, W.L. 1996. Geomorphology and Policy for Restoration of Impounded Rivers: What is Natural? Chapter 18 in The Scientific Nature of Geomorphology, B.L. Rhoads and C.E. Thorn, eds. New York: John Wiley & Sons.

Graf, W.L. 1999. Dam nation: A geographic census of American dams and their large-scale hydrologic impacts. Water Resources Research 35:1305–1311.

Graf, W.L. 2001a. Damage control—Restoring the physical integrity of America's rivers. Presidential Address. Annals of the Association of American Geographers 91:1–27.

Graf, W.L. 2001b. Process reversal for rivers: Fluvial restoration by removal of dams. Abstract of paper to presented at 97th Annual Meeting of the Association of American Geographers, New York, Feb. 28–March 3, 2001.

Graf, W.L., and K. Randall. 1998. A Guidance Document for Monitoring and Assessing the Physical Integrity of Arizona Streams. Arizona Department of Environmental Quality, Contract Report 95-0137, Phoenix, AZ.

Grams, P.E., and J. Schmidt. In Press. Geomorphology of the Green River in the Eastern Uinta Mountains, Colorado and Utah, in Varieties of Fluvial Form. New York: John Wiley & Sons.

Grants Pass Irrigation District. 2001. Press release, July 24. Obtained online at http://www.gpid.com/Consent%20Decree%20-%20Final.htm, November 30, 2001.

Gresch, T., J. Lichatowich, and P. Schoonmaker. 2000. An estimation of historic and current levels of salmon production in the Northeast Pacific ecosystem: Evidence of a nutrient deficit in the freshwater systems of the Pacific Northwest. Fisheries 25(1): 15–21.

Hakanson, L., and M. Jansson. 1983. Principles of Lake Sedimentology. Berlin: Springer-Verlag.

Hall, C.A. 1972. Migration and metabolism in a temperate stream ecosystem. Ecology 53(4): 585–604.

Hanna, S., and S. Jentoft. 1996. Human Use of the Natural Environment: An Overview of Social and Economic Dimensions. Pp. 35–56 in Rights to Nature: Ecological, Economic, Cultural, and Political Principles of Institutions for the Environment. Covelo, CA: Island Press.

Harper, C.L. 2001. Environment and Society: Human Perspectives on Environmental Issues. London: Prentice-Hall.

Hays, S.P. 1959. Conservation and the Gospel of Efficiency: The Progressive Conservation Movement 1890–1920. Cambridge, MA: Harvard University Press.

Heiler, G., T. Hein, F. Schiemer, and G. Bornette. 1995. Hydrological connectivity and flood pulses as the central aspects for the integrity of river-floodplain system. Regulated Rivers: Research and Management 11:351–361.

Heinz Center, The. In press. The State of the Nation's Ecosystems: Measuring the Lands, Waters, and Living Resources of the United States. Cambridge, UK: Cambridge University Press.

Helesic, I., and E. Sedlak. 1995. Downstream effect of impoundments on stoneflies: case study of an epipotamal reach of the Jihlava River, Czech Republic. Regulated Rivers: Research and Management 10:39–49.

Holleman, J. 2001. Don't be fooled: Saluda deadly. The State Newspaper, Columbia, South Carolina, July 24, 2001, p. 1, 6.

Horowitz, A.J. 1991. A Primer on Sediment-Trace Element Chemistry. Chelsea, MI: Lewis Publishers.

Hughes, R.M., and R.F. Noss. 1992. Biological diversity and biological integrity: current concerns for lakes and streams. Fisheries 17(3): 11–19.

Huntington M.W., and J. D. Echeverria. 1991. The American Rivers Outstanding Rivers List. Washington, DC: American Rivers, Inc., p. iii.

Huth, H. 1957. Nature and the American: Three Centuries of Changing Attitudes. Lincoln, NE: University of Nebraska Press.

Iversen, T.M., B. Kronvang, B.L. Madsen, P. Markmann, and M.B. Nielsen. 1993. Re-establishment of Danish streams: Restoration and maintenance measures. Aquatic Conservation: Marine and Freshwater Ecosystems.

Jackson, D.C. 1988. Great American Bridges and Dams. New York: John Wiley & Sons.

Jackson, D.C. 1995. Building the Ultimate Dam: John S. Eastwood & the Control of Water in the West. Lawrence, KS: University of Kansas Press.

Janssen, R., C. Nilsson, M. Dynesius, and E. Anderson. 2000. Effects of rivers regulation on river-margin vegetation: a comparison of eight boreal rivers. Ecological Applications 10(1): 203–224.

Jennings, M.J., L.S. Fore, and J.B. Karr. 1995. Biological monitoring of fish assemblages in the Tennessee Valley reservoirs. Regulated Rivers: Research and Management 11:263–274.

Karr, J.R. 1981. Assessment of biotic integrity using fish communities. Fisheries 6(6): 21–27.

Karr, J.R. 1994. Endangered species: An Overview of Problems and Needs. Seattle, WA: University of Washington.

Kelly, J.R., and M.A. Harwell. 1990. Indicators of ecosystem recovery. Environmental Management 14:527–545.

Kennebec Coalition. 1999. A river reborn: benefits for people and wildlife of the Kennebec River following removal of Edwards Dam. Augusta, ME: Natural Resources Council of Maine.

Kinsolving, A.D., and M.B. Bains. 1993. Fish assemblage recovery along a riverine disturbance gradient. Ecological Applications 3:531–544.

Kline, T.C., J.C. Goering, and R.J. Piorkowski. 1997. The Effects of Salmon Carcasses on Alaskan Freshwaters. Pp. 179–204 in Freshwaters of Alaska: Ecological synthesis. New York: Springer-Verlag.

Knighton, D. 1998. Fluvial Forms and Processes: A New Perspective. London and New York: Arnold and Oxford University Press.

Kondolf, G.M., and M.G. Wolman. 1993. The sizes of salmonid spawning gravels. Water Resources Research 29:2275–2285.

Larkin, G.A., and P.A. Slaney. 1997. Implications of trends in marine derived nutrient influx to South coastal British Columbia salmonid production. Fisheries 22(11): 16–24.

Lee, K.N. 1993. Compass and Gyroscope: Integrating Science and Politics for the Environment. Covelo, CA: Island Press.

Lehmkuhl, D.M. 1972. Change in thermal regime as a cause of reduction of benthic fauna downstream of reservoir. Journal Fisheries Research Board of Canada 29: 1329–1332.

Leopold, L.B., 1994. A View of the River. Cambridge, MA: Harvard University Press.

Leopold, L.B., M.G. Wolman, and J.P. Miller. 1964. Fluvial Processes in Geomorphology. San Francisco: Freeman.

Ligon, F.K., W.E. Dietrich, and W.J. Trush. 1995. Downstream ecological effects of dams. BioScience 45:183–192.

Malanson, G.P. 1993. Riparian Landscapes: Cambridge Studies in Ecology. Cambridge, UK: Cambridge University Press.

Malmqvist, B., and G. Englund. 1996. Effects of hydropower induced flow perturbations on mayfly (Ephemeroptera) richness and abundance in north Swedish river rapids. Hydrobiologia 341:145–158.

Martin, C. 1981. The American Indian as Miscast Ecologist. Pp 243–252 in Ecological Consciousness, R.C. Schulz and J.D. Hughes, eds. Washington, DC: University Press of America.

Marx, W. 1996–97. To the rescue of the southern steelhead. California Coast & Ocean, a publication of the California Coastal Conservancy 12(4): 22.

Matilija Coalition. 2000. Bruce Babbitt Removes Section of 200-foot Matilija Dam on the Ventura River in Southern Califonia. Press release. Matilija Coalition, Ventura, CA.

Mattice, J.S. 1991. Ecological Effects of Hydropower Facilities. Pp. 8.1–8.57 In Hydropower Engineering Handbook. J.S. Gulliver and R. E. Arndt, eds. New York: McGraw Hill Inc.

Meehan, W.R. (editor). 1991. Influences of forests and rangeland management on Salmonid fishes and their habitats. Special Publication 19, Bethesda, MD: American Fisheries Society.

McCullough, D.G. 1968. The Johnstown Flood. New York: Simon and Shuster.

McCully, P. 1996. Silenced Rivers: The Ecology and Politics of Large Dams. London: Zed Books.

McManus, J., and R.W. Duck (editors). 1993. Geomorphology and Sedimentology of Lakes and Reservoirs. New York: John Wiley & Sons.

Miller, J. R., and J. B. Ritter. 1996. An examination of the Rosgen classification of natural rivers. Catena 27:295–299.

Milner, H.H. 1994. System Recovery. Pp. 76–97 in The Rivers Handbook. Oxford, UK: Blackwell Scientific Publications.

Minckley, W. L. 1991. Native Fishes of the Grand Canyon Region: An Obituary? Pp. 124–177 in Colorado River Ecology and Dam Management: Proceedings of a Symposium, May 24–25, 1990, Santa Fe, New Mexico. Washington, DC: National Academy Press.

Murakami, K., and K. Takeishi, 1977. Behavior of heavy metals and PCB's in dredging and treating bottom deposits. Pages 26–42 in S.A. Peterson and K.K. Randolph, eds., Management of bottom sediments containing toxic substances. Proceedings of the 2nd US/Japan Experts meeting, Washington, DC, USEPA-600/3-77-083.

Nash, R. F. 1973. Wilderness and the American Mind. New Haven, CT: Yale University Press.

National Hydropower Association (NHA). 1999. Relicensing Hydroelectric Projects: A Handbook for People Involved in Relicensing Hydropower Projects, Washington, DC: March 1999.

National Park Service. 2001. Wild and Scenic Rivers Act. Obtained online at http://www.nps.gov/rivers/wsract.html, December 17, 2001.

National Research Council. 1983. Safety of Existing Dams: Evaluation and Improvement. Washington, DC: National Academy Press.

National Research Council. 1985. Safety of Dams: Flood and Earthquake Criteria. Washington, DC: National Academy Press.

National Research Council. 1992. Restoration of Aquatic Ecosystems. Washington, DC: National Academy Press.

National Research Council. 1996. Upstream: Salmon and Society in the Pacific Northwest. Washington, DC: National Academy Press.

National Research Council. 1997. Nature and Human Society: The Quest for a Sustainable World. Washington, DC: National Academy Press.

National Research Council. 1999. New Strategies for America's Watersheds. Washington, DC: National Academy Press.

National Sporting Goods Association. 1998. Sports Participation in 1997: Series I and II. Mt. Prospect, IL.

National Weather Service. 1999. Flood Losses: Compilation of Flood Loss Statistics. Obtained online at http://www.nws.noaa.gov/oh/flood_stats/Flood_loss_time_series.htm, August 22, 1999.

Nelson, J.E., and P. Pajak. 1990. Fish habitat restoration following dam removal on a warmwater river. Pp. 57–65 in The Restoration of Midwestern Stream Habitat. American Fisheries Society, North Central Division, Rivers and Streams Technical Committee Symposium Proceedings at the 52nd Midwest Fish and Wildlife Conference, 4–5 December, 1990. Minneapolis, MN.

Newbold, J. D. 1992. Cycles and spirals of nutrients. Pp. 279–408 in The Rivers Handbook: Volume 1 Hydrological and Ecological Principles. London: Blackwell Scientific.

Newcombe, C.P., and D.D. MacDonald. 1991. Effects of suspended sediments on aquatic ecosystems. North American Journal of Fisheries Management 11:72–82.

Nilsson, C., R. Jansson, and U. Zinko. 1997. Long-term responses of river-margin vegetation to water-level regulation. Science 276(2): 798–800.

Odeh, M. (editor). 1999. Advances in Fish Passage Technology: Engineering Design and Biological Evaluation. Bethesda, MD: American Fisheries Society.

Odeh, M. (editor). 2000. Innovations in Fish Passage Technology. Bethesda, MD: American Fisheries Society.

Office of Technology Assessment (OTA). 1995. Fish Passage Technologies: Protection at Hydropower Facilities. Congress of the United States, U.S. Government Printing Office, Washington, DC, OTA-ENV-641.

Olson, B. 2001. Oregon river dam giving way to salmon. Reuters News Service. Obtained online at http://www.enn.com/news/wire-stories/2001/10/10172001/reu_45294.asp, November 30, 2001.

Petts, G. 1984. Impounded Rivers: Perspectives for Ecological Management. New York: John Wiley & Sons.

Petts, G.E. 1980. Long-term consequences of upstream impoundment. Environmental Conservation 7(4): 325–332.

Pisani, D.J. 1996. Water, Land and Law in the West: The Limit of Public Policy 1850–1920. Lawrence, KS: University Press of Kansas.

Poff, N.L., J.D. Allan, M.B. Bain, J.R. Karr, K.L. Prestegaard, B.D. Richter, R.E. Sparks, and J.C. Stromberg. 1997. The natural flow regime: A paradigm for conservation and restoration of river ecosytems. BioScience 47:769–784.

Pohl, M.M. 1999. The Dams of the Elwha River, Washington: Geomorphic Impacts and Policy Implications. Unpublished Ph.D. dissertation, Arizona State University, Tempe.

Pollard, A.I., and T. Reed-Anberson. 2001. La Crosse, WI: North American Benthological Society.

Postel, S.L. 2000. Entering an era of water scarcity: The challenge ahead. Ecological Applications 10:941–948.

Power, M.E., D. Tilman, J.A. Estes, B.A. Menge, W.J. Bond, L.S. Mills, G. Daily, J.C. Castilla, J. Lubchenco, and R.T. Paine. 1996. Challenges in the quest for keystones. BioScience 46:609–620.

Pyle, M.T. 1995. Beyond fish ladders: Dam removal as a strategy for restoring America's rivers. Stanford Environmental Law Journal 14(1): 97.

Rabeni, C.F., and R.B. Jacobson. 1993. The importance of fluvial hydraulics to fish-habitat restoration in low-gradient alluvial streams. Freshwater Biology 29:211–220

Randall, A. 1991. Total and Non-use Values. In J.B. Braden and C.D. Kolstad, eds. in Measuring the Demand for Environmental Quality. Amsterdam: Elsevier Science.

Raymond, H.L. 1988. Effects of hydroelectric development and fisheries enhancement on spring and summer Chinook salmon and steelhead in the Columbia River Basin. North American Journal of Fisheries Management 8(1): 1–248.

Richter, B.D., J.V. Baumgartner, J. Powell, and D.P. Braun. 1996. A method for assessing hydrologic alteration within ecosystems. Conservation Biology 10: 1163–1174.

River Alliance of Wisconsin and Trout Unlimited. 2000. Dam Removal: A Citizen's Guide to Restoring Rivers. Madison, WI: River Alliance of Wisconsin. http://www.wisconsinrivers.org/SmallDams/toolkit-order-info.html.

Rosgen, D.L. 1994. A classification of natural rivers. Catena 22:169–199.

Salomons, W., and U. Forstner. 1984. Metals in the Hydrocycle. Berlin: Springer-Verlag.

Schumm, S.A. 1977. The Fluvial System. New York: John Wiley & Sons.

Schwab, A.K. 2000. Preventing disasters through "hazard mitigation." Popular Government (Spring):3–12.

Scientific Assessment and Strategy Team. 1994. Science for Floodplain Management into the 21st Century. Washington, DC: Interagency Floodplain Management Review Committee, Administration Floodplain Management Task Force. Washington, DC: .Government Printing Office.

Sedell, J.R., G.H. Reeves, F.R. Hauer, J.A. Stanford, and C.P. Hawkins. 1990. Role of refugia in recovery from disturbances: Modern fragmented and disconnected river systems. Environmental Management 14:711–724.

Shuman, J.R. 1995. Environmental considerations for assessing dam removal alternatives for river restoration. Regulated Rivers: Research and Management 11:249–261.

Simons, D. B., and F. Sentürk. 1992. Sediment Transport Technology: Water and Sediment Dynamics. Littleton, CO: Water Resources Publications.

Sittig, M. 1980. Priority Toxic Pollutants: Health Impacts and Allowable Limits. Park Ridge, NJ: Noyes Data Corporation.

Softky, M. 2000. A look at Searsville's past. The Almanac, May 31. http://www.almanacnews.com/morgue/2000/2000_05_31.covside.html.

Solley, W.B., R.R. Pierce, and H.A. Perlman. 1998. Estimated Use of Water in the United States in 1995. Circular 1200. Washington, DC: U.S. Geological Survey.

Solomons, W., and U. Förstner. 1984. Metals in the Hydrocycle. Berlin: Springer-Verlag.

Staggs, M., J. Lyons, and K. Visser. 1995. Habitat restoration following dam removal on the Milwaukee River at West Bend. Pp. 202–203 in Wisconsin's biodiversity as a management issue: A report to Department of Natural Resources managers. Wisconsin Department of Natural Resources.

Stanford, J.A., and F.R. Hauer. 1991. Mitigating the impacts of stream and lake regulation in the Flathead River catchment, Montana, USA: An ecological perspective. Aquatic Conservation: Marine and Freshwater Ecosystems 2:35–63.

Stanford, J.A., and J.V. Ward. 1979. Stream regulation in North America. Pp. 215–236 in The Ecology of Regulated Streams. New York: Plenum Press Publishing.

Stanford, J.A., and J.V. Ward. 1991. Limnology of Lake Powell and the Chemistry of the Colorado River. Pp. 75–103 in Colorado River Ecology and Dam Management: Proceedings of a Symposium, May 24–25, 1990 Santa Fe, New Mexico. Washington, DC: National Academy Press.

State of California. 1997. California's Ocean Resources: An Agenda for the Future. Sacramento, CA: Ocean Resources Management Program.

State of Iowa, Department of Natural Resources. 2002. Iowa Farm Pond Program. Obtained online at http://www.state.ia.us/dnr/organiza/fwb/fish/programs/farmpond.htm, January 4, 2002.

Stevens, L.E., J.P. Shannon, and D.W. Blinn. 1997. Colorado River benthic ecology in Grand Canyon, Arizona, USA: dam, tributary and geomorpholgical influences. Regulated Rivers: Research and Management 13:129–140.

Stromberg, J.C., D.T. Patten, and B.D. Richter. 1991. Flood flows and dynamics of Sonoran riparian forests. Rivers 2:221–235.

Task Committee on Guidelines for Retirement of Dams and Hydroelectric Facilities. 1997. Guidelines for the Retirement of Dams and Hydroelectric Facilities. New York: American Society of Civil Engineers.

Thomas, R.L. 1987. A protocol for the selection of process-oriented remedial options to control in situ sediment contaminants. Hydrobiologia 149:247–258.

de Tocqueville, A. 1850. Democracy in America, Vol. II. P. Bradley, ed. New York: Knopf.

Travnicheck, V.H., M.B. Bain and M.J. Maceina. 1995. Recovery of a warmwater fish assemblage after the initiation of a minimum flow-release downstream from a hydroelectric dam. Transactions of the American Fisheries Society 124:836–844.

Tuan, Y.F. 1974. Topophilia: A Study of Environmental Perceptions, Attitudes, and Values. Englewood Cliffs, NJ: Prentice-Hall.

Twain. M. 1903. Life on the Mississippi. New York: Harper.

Tyus, H.M. 1999. AFS Policy Statement on Effects of Altered Stream Flows on Fishery Resources. http://www.fisheries.org.

U.S. Army Corps of Engineers. 2001. Lower Snake River Juvenile Salmon Migration Feasibility Study Index: Public Information. Obtained online at http://www.nww.usace.army.mil/lsr/, December 12, 2001.

U.S. Bureau of Reclamation. 1980. Safety Evaluation of Existing Dams. Washington, DC: Government Printing Office.

U.S. Bureau of Reclamation. 1987. Design of Small Dams, Third Edition. Highlands Ranch, CO: Water Resource Publications.

U.S. Bureau of Reclamation. 2000. Matilija Dam Removal—Appraisal Report. Denver, CO: Technical Service Center.

U.S. Census Bureau. 1999. Statistical Abstract of the United States: The National Data Book. Washington, DC: U.S. Department of Commerce.

U.S. Department of Energy (USDOE). 1991. Environmental Mitigation at Hydroelectric Projects. Volume I: Current Practices for Instream Flow Needs, Dissolved Oxygen, and Fish Passage. Idaho Falls, ID: Idaho National Engineering Laboratory. Report DOE/ID-10360(V1).

U.S. Department of Energy (USDOE). 1994. Environmental Mitigation at Hydroelectric Projects. Volume II: Benefits and Costs of Fish Passage and Protection. Idaho Falls, ID: Idaho National Engineering Laboratory. Report DOE/ID-10360(V2).

U. S. Department of the Interior. 1996. Elwha River: Ecosystem Restoration Implementation. Draft Environmental Impact Statement, Olympic National Park.

U.S. Department of Transportation. 2001. Domestic Shipping: Vital to the Nation's Economy, Security and Transportation. Obtained online at http://www.marad.dot.gov/publications/PDF/domestic_shipping.htm, October 11, 2001.

U.S. Environmental Protection Agency (EPA). 1989. Report to Congress: Dam Water Quality. Office of Water Regulations and Standards, Washington, DC. EPA 506/2-89/002.

U.S. Environmental Protection Agency (EPA). 2001. National Management of Measures to Protect and Restore Wetlands and Riparian Areas for the Abatement of Nonpoint Source Pollution. Draft Report. Washington, DC: Office of Water. EPA 841-B-01-001.

U.S. Federal Inter-Agency River Basin Committee, Subcommittee on Benefits and Costs. 1950. Proposed Practices for Economic Analysis of River Basin Projects. Washington, DC: Interagency Commission on Water Resources.

U.S. Geological Survey. 2000. The Quality of Our Nation's Waters. Circular 1225. Washington, DC: U.S. Geological Survey.

Vannote, R.L., G.W. Minshall, K.W. Cummins, J.R. Sedell, and C.E. Cushing. 1980. The river continuum concept. Canadian Journal of Fisheries and Aquatic Sciences 37:130–137.

Ventura County Flood Control District. 2001. Matilija Dam Ecosystem Restoration Feasibility Study. Obtained online at http://www.matilijadam.org/index.htm, December 13, 2001.

Vinson, M.R. 2001. Long-term dynamics of an invertebrate assemblage downstream from a large dam. Ecological Applications 11(3): 711–730.

Voelz, N.J., and J.V. Ward. 1991. Biotic responses along the recovery gradient of a regulated stream. Canadian Journal of Fisheries and Aquatic Sciences 48: 2477–2490.

Wahl, K.L., W.O. Thomas, and R.M. Hirsch. 1995. Stream-Gaging Program of the U.S. Geological Survey. Circular 1123. Denver, CO: U.S. Geological Survey.

Waldman, C. 1985. Atlas of the North American Indian. New York: Facts on File Publications.

Walling, D.E., and B.W. Webb. 1996. Water Quality I: Physical Characteristics. River Flows and Channel Forms. London: Blackwell Science.

Ward, J.V. and J.A. Stanford. 1979. Ecological factors controlling stream zoobenthose with emphasis on thermal modifications of regulated streams. Pages 35–55 in J.V. Ward and J.A. Stanford, eds. The Ecology of Regulated Streams. New York: Plenum Press.

Ward, J.V., and J.A. Stanford. 1995. Ecological connectivity in alluvial river ecosystems and its disruption by flow regulation. Regulated Rivers: Research and Management 11:105–119.

Waterwatch. 2001. Remove Savage Rapids Dam: Saving Salmon. Obtained online at http://www.waterwatch.org/savagerapids/savage_rapids_dam4.html, May 17, 2001.

Webb, B.W., and D.E. Walling. 1996. Water Quality II: Chemical Characteristics. River Flows and Channel Forms. London: Blackwell Science.

Webster's New Collegiate Dictionary. 1981. Springfield, MA: G.&C. Merriam Co.

Wegner, D.L. 2000. Looking toward the future: The time has come to restore Glen Canyon. Arizona Law Review 42(2): 239–257.

Weitkamp, D.E., and M. Katz, 1980. A review of dissolved gas supersaturation literature. Transactions of the American Fisheries Society 109:659–702.

White, R. 2001. Evacuation of Sediments from Reservoirs. London, UK: Thomas Telford Ltd. Distributed by ASCE Press, Reston, VA.

Williams, D.D., and H.B.N. Hynes. 1976. The recolonization mechanisms of stream benthos. Oikos 27:265–272.

Williams, G.P. 1978a. Bank-full discharge of rivers. Water Resources Research 14:1141–1154.

Williams, G.P. 1978b. The Case of the Shrinking Channels of the North Platte and Platte Rivers in Nebraska. Circular 781. Washington, DC: U.S. Geological Survey.

Williams, G.P., and M.G. Wolman. 1984. Downstream effects of dams on alluvial rivers. Professional Paper 1286. Washington, DC: U.S. Geological Survey.

Williams, J.E., C.A. Wood, and M.P. Dombeck (editors). 1997. Watershed Restoration: Principles and Practices. Bethesda, MD: American Fisheries Society.

Willson, M.F., S.M. Gende, and B.H. Marston. 1988. Fishes and the forest: Expanding perspectives on fish-wildlife interactions. BioScience 48(6): 455–462.

Wik, S.J. 1995. Reservoir drawdown: Case study in flow changes to potentially improve fisheries. Journal of Energy Engineering 121(92): 89–96.

Winter, B.D, 1990. A brief review of dam removal efforts in Washington, Oregon, Idaho and California. U.S. Department of Commerce, NOAA Tech. Memo. NMFS F/NWR-28.

Wohl, E. 2000. Mountain Rivers. Washington, DC: American Geophysical Union.

Wood, P.J., and P.D. Armitage, 1997. Biological effects of fine sediment in the lotic environment. Environmental Management 21(2): 203–217.

World Commission on Dams. 2000. Dams and Development. Herndon, VA: Stylus Publishing. Also available online at http://www.dams.org/report.

Yalin, M.S. 1992. River Mechanics. New York: Pergamon Press.